AND FROM HERE ON...

BY
SOURJYA GUPTA

ISBN 978-93-5458-953-9

© Sourjya Gupta 2021

Published in India 2021 by Pencil

A brand of

One Point Six Technologies Pvt. Ltd.

123, Building J2, Shram Seva Premises,

Wadala Truck Terminal, Wadala (E)

Mumbai 400037, Maharashtra, INDIA

E connect@thepencilapp.com

W www.thepencilapp.com

DISCLAIMER: *The opinions expressed in this book are those of the authors and do not purport to reflect the views of the Publisher.*

AUTHOR BIOGRAPHY

Sourjya is a young man with varied interests; he dabbles in painting and sketching, plays the guitar and violin and has also written a few poems. Sourjya has tried to present his wonder about the unimaginable expanse of space, the mind-boggling variety of life, the mysteries of the human body and the logic of mathematics, for his young colleagues. 'And From Here On …' is Sourjya's homage to those great scientists and thinkers through the ages, who have shaped our current understanding of life and our universe!

TO MY
GRANDPARENTS

"LITTLE SCIENCE TAKES YOU
AWAY FROM GOD BUT MORE
OF IT TAKES YOU TOWARDS
HIM."... LOUIS PASTEUR

FOREWORD

I have known Sourjya since his childhood as a poet and as an artist whose paintings had clarity and character. But exploring science in a manner which classroom teaching does not always permit, with a spirit of inquiry which is beyond requirements of curricula has always been an integral part of his thinking. In a sense, the book 'And From Here On ...' was waiting to happen. But it is the expanse of the book that is breath-taking; something I was not prepared for. It is amazing that young Sourjya, a high school student, could conceive of and actually string together a narrative of this range, flow and analysis in such a short span of time. The book cannot be classified easily; nor is it important to classify it. It is broadly about connecting the evolution of scientific theories, in a course parallel to human civilization. But its uniqueness lies in the way the stages are connected and the scientific theories elaborated. It reads more like natural philosophy, bringing one back to an era when knowledge was not stratified and barriers were not built around academic enquiries. I am told that the idea of the book evolved when Sourjya would take long walks with his mother during the lockdown and describe the sequence as it evolved in his mind. Later encouraged by his parents and sister, he wrote this down in the form of a book. Was it that straightforward? Not quite – we see the glimpses of a budding renaissance scholar. Of course, our young author modestly admits that in the context of the "ocean of knowledge", the focus is on the "drops"— the specifics of observations that we encounter in our everyday lives. He then explores these, connecting them to the theories of science. Surely it is a drop that maps the ocean! The book takes across different disciplines within natural sciences. It opens with Lamarck's theory and goes to the theory of Evolution by Charles Darwin. Then in a gorgeous sweep, it comes to the Drake equation of Fermi, which provides the probability of finding intelligent beings in our galaxy to the second law of thermodynamics proposed by the German mathematician and physicist Rudolf Julius Emmanuel Clausius. The book is divided into smart subtitles which enables the reader to understand the connectivity of the body of the text. From the possibility of life in the Milky Way, the author, takes us to the Doorway of destruction, which starts with the discovery of isotopes by Otto Hahn, the German Chemist. It then moves to the analysis of energy released in explosion and its source. The book takes us across aeons, across dimensions — sub atomic to the galactic and across the range that covers the fundamentals and the frontiers in lucid sweeps. I have read the book as a lay reader who did not study science formally beyond high school and found it to be unputdownable. The book is a wonderful tribute to the scientific community and people from all professions who practice science for the benefit of mankind. But it is an invaluable gift for young students; contemporaries of Sourjya and also the entire community of readers who are interested in knowledge as a way of life. I am sure it will build scientific temperament and evoke enquiries beyond class-room learning. I compliment the young author Sourjya for the book. It has immense possibilities of being made into an E book, an audiobook and finally a movie!

Anita Agnihotri.

Indian Bengali Author and former Civil Servant.

Kolkata,

August, 2021

ACKNOWLEDGEMENTS

I was always intrigued by science; especially the way human knowledge expanded through the ages to make us what we are today. I did a lot of reading outside school textbooks. What astonished me were the links between the various concepts and developments through the centuries. They fitted into a giant picture of our present-day life like pieces in an enormous jigsaw puzzle. It was during my evening walks with my mother, that I spoke of what I had read and how interconnected everything was. My mother, an avid student of science would egg me on, probing deeper and deeper. My father encouraged me to structure my thoughts into the form of this book. He has sat with me through long nights, guiding me through the process. This book is actually a written account of all that I shared with my mother during those evening walks in the midst of the pandemic induced lockdowns! That this book is literally in black and white is solely due to the encouragement of my parents and my sister. They also helped enormously by hearing out chapters as I wrote helping me improve the content continually. My father and sister went through the text and further improved it. I am also grateful to Dr. Satyajit

Ash whose influence during the last couple of years has played an important role in this effort of mine. I would also like to thank Pencil Publishers who guided me all through, till the publication of this book.

EPIGRAPH

Not only is the universe stranger than we imagine, it is stranger than we can imagine ...

Arthur Eddington

CONTENTS

A DROP FROM THE OCEAN

Have you ever wondered why the world is growing? Well, I guess it is because most people don't apply their minds enough when something happens before their eyes incessantly. Rather they continue to do what they are programmed to do. If they did so, the world would have been a stagnant land of philosophers. The development we have seen so far is a product of not only the thinkers but also the farmers, artisans, factory workers and workmen who are skilled in their field, along with the doctors, engineers, economists and lawyers. Before exploring the great minds and their works, we should appreciate all these people, the practitioners who apply the scientific knowledge and power for the benefit of humankind.

The great scientist Isaac Newton once famously remarked, what we know is a drop, and what is still to be known is an ocean! Let us keep this ocean away from our minds for now and concentrate on the drop. In the following sections, I would be covering only a fraction of the drop that we encounter in our everyday life, and on special occasions. I hope that the exploration of these fragments, these tiny droplets, would interest many.

Let us go back in time, when the drop just began to form, and take shape. Tools, such as the hammer and knives, were invented long back by our very great ancestors. These were probably some of the earliest human inventions. But were they really invented by humans? *Australopithecus afarensis,* one of the earliest species linked to humans, are known to have made and used tools carved from stone. These *Australopithecus* have their common forefathers with chimpanzees called *Homini.* Did a long-distance cousin of the forefathers of the present-day chimpanzees make our first tools then? We humans, *the Homo sapiens* that we are, are definitely more complex and far more "accomplished" than a banana peeling chimpanzee! What is this force, this magic, which has made us more and more intricate and more and more intelligent? If our ancestors were cousins of chimpanzees, and if we are continuously becoming more complex over the last millions of years, then definitely at some point in history, we were like a block of jelly, like any life in the least complicated form.

This fact, though difficult to believe, is true! May be this does not surprise us as we have been hearing this since we were children. But imagine your grandfather's grandfather and his grandfather a million times over, and we reach a blob of jelly!

What is the magic that has caused us to develop into the complex beings that we are today from our (great)^N grandfathers, the blobs of jelly? A reasonable theory was propounded by one of the earliest prominent evolutionary scientists, Jean Baptiste Pierre Antoine de Monet Chevalier de Lamarck. Lamarck's theory is based on a vast analysis, but we may take note of the two laws proposed by Lamarck. First, the law of use and disuse, which states that the part of the body which is used more and more gets strengthened, and those with little use lose their strength and eventually disappear or remain as a vestigial organ. This is why monkeys, which live on trees and have a need to grip branches have tails, while chimpanzees, which live on the ground (as we humans do, as their descendants) do not have tails. The second law says that these modifications of organs pass to the subsequent generations through reproduction.

It is now evident that a new subject was emerging on the horizon of human knowledge. On 12th February 1809 was born the father of evolutionary theory, Charles Robert Darwin in Shrewsbury in Shropshire in England. Nearly fifty years of hard work, educational learning, thinking, discovering and travelling, with friends and colleagues resulted in one of the most significant scientific publications of all time, "Evolution by Natural

Selection", which centred around variation, reproduction and heritability. British naturalist and explorer Alfred Russel Wallace independently proposed the theory of evolution through natural selection. His writings on the subject were published along with some of Darwin's writings. The Natural Selection theory, immediately after its publication attracted the world's attention. Darwin's theory led to international debates and shook the very foundations of religious ideologies of the time. Darwin's magnum opus Origin of Species, published on 24th November 1859, is accepted to be the foundation of evolutionary biology. After reading the Origin of Species, Herbert Spencer, another early evolutionary scientist, propounded the popular concept of "Survival of the fittest". What is the principal element of this theory? Darwin noticed that species reproduce more than their capacity to sustain in their habitat. But why would they do so? An example would help us understand his concept.

Let us imagine a patch of forest with a few giraffes and trees. If in that patch, every couple of giraffes gives birth to 10 offspring, then within a few generations, the world will be filled with giraffes! But this never happens in reality. This never happens as Darwin's natural selection comes into play. Every offspring from a couple of giraffes will have minor differences with its siblings. Some will be short, some tall, some strong, some weak. Let us say they reside in a forest with tall plants, which is only accessible to the taller giraffes. The taller

> *The fittest in a species in one environment may become less fit with change in circumstance. The mammoth was most fit in the most recent ice age but with warming of the climate and hunting by humans, it became less fit and extinct a few thousand years later.*

giraffes can satisfy their hunger by munching the leaves of the tall plants, and the shorter ones, starved of food, weaken and die and will not reproduce. So, the next generation coming from the taller giraffes will carry this trait; that is, they would be taller. If this process continues for long over several generations, significant differences can occur. This is just one example; many factors have led the animals and plants to evolve. The organisms which are best adapted to their environment, in other words, those which are fittest in the respective species survived and reproduced, therefore "survival of the fittest".

Evolution can affect any aspect of the species. If humans evolved in Asia, they could evolve with certain traits based on their environment. On a very similar note, humans living in South America would evolve, but it is unlikely with little probability that two similar beings evolve from different ancestors at different geographical locations as environments would differ. This problem was a little disturbing; in a sense, as it implied that humans travelled on a ship and spread their civilization many thousands of years before they knew the technology. As a result, humans who evolved on different continents were similar. But what about the other animals who did not travel by ship? Even if we assumed animals used to fly back then (which is not true), what about the plants? There is a fern called *Glossopteris* whose fossils were found in Australia, Antarctica, India, Africa and South America, many thousands of kilometres apart! Yet, these plants and animals are found to be broadly similar. So, what is going on exactly?

The German geophysicist and meteorologist, Alfred Lothar Wegener came up with an ingenious solution to this puzzle. While he was playing a jigsaw puzzle with continents on a map, he noticed something amazing and something that had eluded the human mind for centuries. He noticed how South America fitted like a key in a lock with Africa and North America with Europe. His exploration revealed that their coastlines have similar rocks. He

Wegener's theory during the 1930s was rejected by most geologists; it was re-established as a part of the theory of plate tectonics during the 1960s. [Britannica]

proposed that like the animals which evolved, so did the map of Earth! According to Wegener, the world was once a single landmass; he named it Pangea (Greek word for unified land). And then the continents slowly drifted apart. Though his continental drift theory was controversial back then and led to huge debates, now we have evidence supporting it. Let us go back to the problem of the fern *Glossopteris* and also many other animals and plants of that epoch. No, they did not need to travel overseas in sailboats! Instead, they shared a common home, the Pangea. It is estimated today that the Americas and Africa and Europe are drifting apart at the rate of about 2.5 cm a year. Now we know about continental drift, and we will know, millions of years later, the world map will probably not be like as we know it today.

Knock, knock! Anyone else there?

So far, so good. We have seen the species with maximum complexity and highest intelligence, the *Homo sapiens*. But that's on Earth. Humans were now knowledgeable enough to guess, Earth is not the only habitable planet in our great universe. Their curiosity began to grow; they began to argue that if there are other habitable planets, then evolution must have occurred there too! Such planets must harbour diverse life forms from the simplest unicellular bacteria to animals as complex and evolved as us. Aliens, as we call them, have been an integral part of sci-fi movies and stories, nearly everyone's childhood fantasy. Though aliens seem more unreal to us, back then, the scientific community started to realize it's a matter of great concern. You must have heard about the father of the nuclear age, the Italian physicist Enrico Fermi. Though he is known for the concept of the nuclear reactor, he came up with an equally interesting concept, now better understood by an equation called the Drake equation named after the American astrophysicist Frank Donald Drake. Fermi asked the question "Where is everybody?". He wondered why no extraterrestrial civilizations had been detected despite the great size and age of the universe.

The Drake equation provides the probability of finding intelligent beings in our galaxy, the Milky Way. It is conjectured that billions and billions of stars in our galaxy have planets with potential life on them. Also, there are billions and billions of galaxies in our observable universe. These

> *Frank Drake, from his equation, estimates 1000-100000000 planets with civilization is in our galaxy Milky Way. According to an estimate, there are 2 trillion galaxies in the observable universe.*

huge numbers indicate that there is a possibility of many planets that could be home to intelligent beings. The numbers of such planets are so huge that it should be certain that we communicate with at least one such civilization. But the fact is that no such communication has been made so far. The first thing that comes to mind is that it is only a probability, so nothing is wrong in the whole situation, and probability is only a possibility and not a certainty. Actually, yes, nothing is wrong. But let us view it from another angle.

Imagine a glass of hot water and a glass of cold water are mixed in a bucket. What's more likely, it will come to a temperature in between, or it would turn hotter or colder? From common sense, it is not expected that hot water gets hotter and cold water gets colder. It was proposed by the great German mathematician and physicist Rudolf Julius Emmanuel Clausius; it became a law of physics, the second law of thermodynamics. But is it theoretically impossible? Let us see at a microscopic level: the fast-moving water molecules constitute the heat, and the slow ones the cold. Now when brought together and a fast molecule hits a slow one, its own speed decreases while increasing that of the other. However, in rare cases, depending on the direction of the collision, it is possible that the speed of the faster one increases further and that of the slower one decreases further. It is not one or two particles in reality. There are billions and billions of particles, making it very unlikely that most of the hot ones get hotter and cold ones get colder. It is much more likely to happen the other way round, that the hot ones slow down and get colder and the cold ones speed up and get hotter, and that is what is observed. So is the likelihood of our finding of aliens, which has a much higher probability than not finding them. Here the unlikely event has happened; we have not found them yet, and that's what is enthralling our minds. This so very unlikely event happening came to be known as the Fermi paradox.

A DOORWAY TO DESTRUCTION

Now when we have become a little familiar with Fermi, let us have a quick look at his most notable discovery, the nuclear reactor. The German chemist, Otto Hahn, with his colleague Austrian physicist Lise Meitner, discovered radioactive isotopes of many elements. Hahn, with his other colleague, German chemist, Fritz Strassmann discovered the nuclear fission reaction, which was explained theoretically by Lise Meitner and her nephew Otto Robert Frisch. These developments attracted Fermi's attention. While scientists began investigating the fission reactions, Fermi started thinking about a self-sustaining, radioactive, chain fission reaction where neutrons produced by the fission of one atom could lead to the fission of other atoms. He discovered that slow neutron bombardments were more effective than fast ones and gave more neutrons from radioactive decay, further bombarding other atoms, growing exponentially. He made the world's first nuclear reactor. American physicist, Julius Robert Oppenheimer, the Director of the Los Alamos Laboratory considered the father of the atom bomb, got a great idea with Fermi's discovery. The Manhattan Project was a research and development project led by the United States and supported by the United Kingdom launched during World War II for developing nuclear weapons. Many prominent scientists worked on the Manhattan Project in the Los Alamos Laboratory, trying to find the limits of explosive power, finally succeeding in making the first atomic bomb. The first atom bomb was tested on 16th July 1945 in a remote desert in Alamogordo Bombing and Gunnery Range, New Mexico, yielding energy equivalent to a blast of 22 kilotons of TNT (trinitrotoluene), much more than expected. While witnessing the explosion Oppenheimer recited a stanza from the sacred text Bhagwat Gita "I am Death, the destroyer of worlds" (*Kaalo asmi loka kshaya kritpraviddho*)". But unfortunately, this massive human achievement was discovered at the wrong time. After the Second World War ended, the US government employed this discovery, committing one of the most

The Manhattan Project employed 129,000 workers and cost nearly $2 billion in those days (equivalent to about $23 billion in 2019).

hideous crimes against humanity. They deployed two such bombs in Hiroshima and Nagasaki in Japan, causing the death of nearly a lakh of civilians and leaving lakhs scarred for life and generations. Oppenheimer and other scientists were extremely disappointed by this misuse of science.

> *1 kiloton of TNT is equal to 4,184,000,000,000 Joules of energy. This is enough to bring 13,000,000 liters of water or five Olympic-size swimming pools at room temperature to boiling!*

When nationalists in different countries realized the immense destructive power of the atom bomb, they began their competition for making bigger and more destructive atom bombs. With new technology and tweaking of the design of the atom bomb, American nuclear physicist, Edward Teller with his team developed the world's first thermonuclear bomb, the hydrogen bomb, the Ivy Mike. This is a specialized atom bomb that works on the principle of fusion and not fission like its' predecessor. To demonstrate their national pride, the USA and Russia have tested nearly a thousand atom bombs since then, significantly disturbing the Earth's natural environment. Fortunately, the spate of testing came to an end with the test of one such Hydrogen bomb developed by Russia, the Tsar Bomba. The Tsar Bomba had 1570 times more yield than the Little Boy and Fat Man dropped in Japan. Tsar Bomba's explosion caused an earthquake measuring nearly 5.0 on the Richter Scale, with the debris reaching a height of about 60 kilometres above the Earth's surface, escaping the Earth's atmosphere and reaching space! The biggest explosion ever conducted!

THE ENERGY EMBEDDED

So big an explosion yields an immense amount of energy; fission reactions in nuclear reactors generate power. Where is this power that is released either during explosions or in reactors, stored? We know, in thermal power plants, the ignition of the coal breaks down the complex molecular structures and forms new bonds in carbon dioxide, releasing energy as heat. But what about radioactive uranium used in nuclear plants? It may be common knowledge that here, we do not break bonds of molecules; we 'break bonds inside atoms' so to speak, causing nuclear decay. The mass of the products is found to be less than the original atom. This mass converts into a huge quantum of energy through mass-energy equivalence, to which we shall come back later.

Let us take the example of uranium-235. This particular isotope of uranium has 92 protons and 143 neutrons in its nucleus. There can be many radioactive reactions; we will consider one. Let us say a slow-moving neutron hits the nucleus of uranium-235, making it unstable uranium-236. Being unstable, the nucleus breaks and one possible yield can be krypton-89 and barium-144 with three individual neutrons and energy in the form of gamma rays. Let us check if the numbers add up, barium has 56 protons, and krypton has 36 protons. They add up to a total of 92 protons, the same as the number that was in the uranium nucleus. We can thus conclude, no proton is lost. Let us check for neutrons; there are 53 neutrons in krypton and 88 neutrons in barium. The total number of neutrons in both the elements add up to 141, and there are three individual neutrons. Uranium-235 had 143 neutrons but was bombarded with one more neutron making a total of 144. We, therefore, conclude that neutron is not lost. But wait, where did the energy come from then?

Well, we know some mass was converted into energy from mass-energy equivalence. We also know of Lavoisier's law of conservation of mass and many laws of conservation of energy. Before we examine this question further, let us know the fundamentals of these laws. The laws resulted from long and painstaking research and were monumental achievements by many scientists and philosophers through the ages. Galileo, the Italian

> *Leibniz entered in the university at the age of 14 and completed his Bachelor's degree at the age of 15 and his Master's degree at the age of 17! By the age of 19, he earned a license to practice law and a Doctorate in Law!*

astronomer, demonstrated that the potential energy of an oscillating pendulum changes to kinetic energy and vice versa as long as it swings. Huygens, the great Dutch scientist and philosopher, made the law for collisions. Gottfried Wilhelm Leibnitz, the German mathematician, propounded a theory on kinetic energy, apart from his very well known contribution to modern calculus. His work influenced Emilie du Chatelet, a French natural philosopher. She translated Newton's famous tome *Philosophy Naturalis Principia Mathematica* to French. Inspired by this book, she publicized an experiment devised by Willem Jacob's Gravesande, a Dutch mathematician and philosopher, which finally led her to propound the law of conservation of energy. With the further addition of friction, heat and work by James Prescott Joule, an English physicist and mathematician, this became the First Law of Thermodynamics.

We are back to the problem. We know that whatever missing mass was there was converted to energy, neither mass got destroyed, nor energy got created; they just got converted, obeying the above laws.

Now we have crossed the first part. To answer the question as to how the energy was released, let us recap a little science from our school days. Long long ago, in

India, Maharishi Kanada and in Greece, Democritus, philosophers, were contemplating what was the smallest size to which something could be cut. No technology back then existed to explore how small anything could be cut to. So they theorized the smallest part, "*anu*" for Kanada and "*atomos*" for Democritus. The Greek word *atomos* means "cannot be divided further". Their quest was never forgotten but not answered clearly until the famous English chemist and physicist John Dalton, propounded the atomic theory. Though among his five postulates which constitute the atomic theory, only two hold true till this day, his theory opened a field in science. The first postulate that atoms are indivisible does not hold true for obvious reasons. The second postulate that all atoms of a specific element have the same mass also does not hold true, as much after Dalton's prediction, Joseph John Thomson, another British physicist, discovered isotopes. One last error was in postulating that atoms cannot be divided into smaller particles. Which of his postulates did he get right? Before going into the last two postulates, Dalton made a significant contribution to our understanding of the nature of atoms. Though some of his steps were wrong, they did provide clues to the right ones to other scientists who succeeded him.

Dalton's last two postulates state that atoms can be rearranged, combined, or

separated in chemical reactions, and they combine with other atoms in whole-number ratios, forming molecules.

Apart from his theory on colour blindness, the conceiving of the atomic theory was an outstanding achievement. The atomic theory was taking birth. French engineer Charles-Augustin de Coulomb proposed the Coulomb's law which gives a quantitative value of force between electric charges. This force was recognized at the macroscopic scale. It was seen that the law holds true microscopically as well. But what are these microscopic charges? It was in 1897 that the British physicist, Joseph John Thomson cracked the mystery of charge with his discovery of the fundamental particle constituting the negative charge, the electron. According to the Coulomb's law, two negative or two positive charges repel each other and that negative and positive charges attract each other.

J J Thomson found that the atom consisted of smaller parts, demolishing one of Dalton's postulates.

Thomson's discovery of the electron gave rise to a mystery; if atoms were made of electrons, why were they electrically neutral? Why were atoms not constantly discharging? Thomson predicted there is something positively charged in the atom too!

Eugen Goldstein had discovered positively charged hydrogen long before Thomson's atomic model. But it required a scientist from New Zealand to travel all the way to Cambridge, where the research was going on, to unravel the mystery. It was none other than Ernest Rutherford. He identified that the positively charged hydrogen was the neutralizing 'thing' in Thomson's model of the atom. Rutherford named it proton. Finally, the anatomy of the atom was taking shape. It was then believed that atoms were like balls consisting of an evenly spread equal number of protons and electrons. This model of the atom came to be known as the plum pudding model after a similar-looking popular English dessert.

Although the plum pudding model explained the electrical neutrality of the atom, Rutherford was still not satisfied; he kept on experimenting with the phenomenon of radioactivity. One of his experiments was the famous gold foil experiment. In that experiment, positively charged alpha particles were shot at a fine gold foil, and the deflection of the particles was recorded on a circular screen beyond the foil. If the electrons and protons were evenly distributed in the atom, as was proposed in the plum pudding model, the alpha particles were not expected to deviate much from their original trajectory. But observation showed huge deviations, and some of the particles were deflected right back! This observation led Rutherford to conclude that there was some massive positive charge in the atoms concentrated in a small space. Soon neutrons were discovered by the British physicist, James Chadwick. The Danish physicist, Neils Bohr finally stabilized the idea of the atom by concluding that, the positively charged protons and electrically neutral neutrons were located in the centre. And far away from this nucleus of protons and neutrons, electrons orbited the nucleus much like the planets revolving around the sun! And there is nothing in the space between the nucleus and the revolving electrons. An astonishing fact became evident that contrary to what we see around us, we are, almost completely vacuum! If the vacuum is hypothetically removed, then the whole Earth would collapse to a size of a sugar cube!

But wait, are we missing something? We saw that Coulomb's law holds true even at the microscopic level.

Then how is it, that protons, all with a positive charge, are huddled together in the centre of the atom? There are neutrons, too, but they do not experience Coulomb repulsion because they are not charges, that is they are neutral. But should the protons not face a colossal force of repulsion? Is it possible that the gravitational attraction between the protons hold them together? No, because the electric force that repels the protons is nearly a trillion, trillion, trillion times greater than the gravitational force of attraction between them, as calculated!

This mystery was unravelled by the Japanese physicist Hideki Yukawa when he discovered a new fundamental force in nature; a strong nuclear force. It is an unbelievably strong force; the force between two protons in the helium nucleus would be equivalent to an average hand resting on someone's lap! At the microscopic level, this force is enormous.

But the discovery of this force leads us to a rather troubling question! If indeed the strong nuclear force is so strong, why is it that we are not collapsing into a supermassive body? The answer lies in the inherent property of this strong nuclear force, which acts over very small distances. If the distance between two protons is increased, the strong nuclear force decreases dramatically, and the electrostatic repulsive force surpasses the strong nuclear force.

Finally, we have all the tools to crack the original mystery, what happened to the uranium-235 atom?

Let us playback the whole scene. A uranium-235 atom is bombarded by a solitary neutron converting it to uranium-236. The size of the atom increases, and so does the diameter of the nucleus of uranium. The increase in the size of the nucleus leads to a decrease in the strong nuclear force holding the nucleus together. The decrease in this force leads to the splitting of the atom into barium-144 and krypton-89. Now the strong nuclear force has to hold two smaller nuclei together, its potential therefore decreases. The potential, due to the strong nuclear force, was so condensed in such a tiny atom that it could be considered mass. Mass energy equivalence essentially states matter is condensed energy; the decrease in potential gives us a clue to the missing mass. This mass got multiplied by the square of the speed of light, releasing a tremendous amount of energy!

THE FORCE UNSEEN

Why is Strong Nuclear Force called so? You must have guessed it! There must be something weak! It is Fermi again. Enrico Fermi theorized the weak nuclear force, another fundamental force of nature. What does this weak nuclear force do, then? So far, we have known we can break atomic bonds in a molecule, giving rise to new molecules. We have also seen that we can break down a nucleus by bombardment with other particles, leading to the separation of a part of protons and neutrons from the mother nucleus, thereby giving birth to new elements. Can we go on doing this, breaking up indefinitely? Or is there a limit to the extent we can break? Dalton limited it to the atom, but eventually, through the efforts of many scientists, we arrived at an understanding of the nature of the atom, with the matter concentrated in the nucleus. The French engineer and physicist, Antoine Henry Becquerel, the grandson of Antoine Cesar Becquerel, one of the pioneers who studied electricity and the phenomenon of luminescence, was the son of Alexandre-Edmond Becquerel, the discoverer of the photovoltaic effect; he was determined to continue the family's hold on physics! He discovered spontaneous radioactivity. This was followed up by extensive research on it by Wilhelm Conrad Roentgen and the Curies. With pioneering advances by Marie and Pierre Curie, we came close to the possibility of breaking the nucleus. But soon, they realized that something else is going on as well. Fission and fusion involved some agents causing them.

Before we move on, just some brief information on the Curies, the most prominent being Marie and Pierre Curie. Their research on radioactivity that unravels the mystery also has extensive contributions, from their daughter Irene-Joliot Curie and her husband, *The Curies discovered thorium is radioactive and then identified the two new radioactive elements, polonium (in honour of Marie Curie's motherland) and radium. From a tonne of pitchblende, an uranium ore, they could separate one-tenth gram of radium chloride!* Frederic-Joliot Curie. And it seems that science was in the genes; the Curie family was a descendant from another famous Physics family – the Bernoulli family.

We go back to their discoveries of the spontaneous decay of radioactive elements. The three most prevalent types of decay are gamma, beta, alpha. Earnest Rutherford was studying these decays. It was concluded gamma decay is associated with the emission of gamma rays which is an electromagnetic wave. Alpha decay is associated with the release of alpha particles which is helium-4 with no electrons, that is, the helium nucleus, thus possessing a +2 charge.

We are left with beta decay. It so happens that the beta decay helps to solve the mystery of the weak nuclear force. It is due to this weak nuclear force something happens, not on a molecular scale, not even at the atomic scale, but at the level of the sub-atomic particle. Beta-decay, as discovered by Rutherford, was accompanied by the release of an electron. Where did this electron come from? It was not one among those orbiting the nucleus because it was observed to be a nuclear phenomenon.

It is a neutron that emitted the electron, with another particle having no charge, called the anti-neutrino. This was caused by the weak nuclear force. What happened to the neutron then? It converted to a proton. This led physicists to conclude that subatomic particles can change among themselves! This came to be known as beta minus decay.

Beta plus was another beta decay discovered later, which explains how we get the neutron back. Neutron initially released an electron and anti-neutrino. Assuming nobody is ready to give the proton (generated from the neutron) back an electron, the proton has to find another way. Getting a negative electron is equivalent to getting rid of a "positive electron". It achieves the objective with anti-neutrino. It cannot absorb anti-neutrino because nobody is ready to give that, so the proton emits a neutrino (which possesses a unit positive charge), converting it back to neutron! Is it not fundamentally amazing how protons and neutrons are interchangeable? But in the process, we found something new. How can an electron be positive? For now, let us accept it; we shall discuss the theory soon. We will call this positive electron – positron. Familiar with the name? Yes, this is the crucial ingredient in the PET (positron emission tomography) scan, one of the most potent scans used to detect abnormalities inside the human body: a fantastic invention by a team led by Edward Joseph Hoffman, one of the pioneers. In the PET scan, a biologically active molecule containing a short-lived radioactive tracer isotope is injected into the body, usually in the blood stream. These molecules accumulate in the tissues of interest. The molecules undergo positron-emission decay (positive beta decay), emitting a positron each. These positrons travel through the tissue (typically less than 1 mm) losing sufficient kinetic energy to enable them to interact with electrons. Each positron encountering an electron results in the annihilation of both, with the release of two gamma photons moving in approximately opposite directions. These are detected by sensors to produce the images which are then analyzed.

This amazing machine can track activities in your body on a real-time basis. That means that if you are playing the guitar, the PET machine can track what part of your brain is activated.

UNIFICATION OF FORCES

Classical physics had known three forces, gravity, electric and magnetic. With the painstaking efforts of many scientists, about which we will learn later, electric and magnetic forces were found to be fundamentally the same; that is, they are interdependent. Scientists were intrigued by two more forces coming into play, the weak nuclear and strong nuclear forces. They wanted to know the source of all these forces. Through the research of Pakistani theoretical physicist, Mohammad Abdus Salam, and Americal theoretical physicists Steven Weinberg and Sheldon Lee Glasow, they were able to unite the strong and weak nuclear forces and also the weak nuclear force to the electromagnetic force. So man finally brought all forces together, well almost! Gravity remained as a sore thumb. The Grand Unified Theory (GUT) is a model in particle physics where at high energies, electromagnetic, weak and strong forces are merged into a single force. Two such prominent theories exist, the gravitational loop theory, with major contributions from Abhay Vasant Ashtekar and string theory, with major contributions from Gabriele Veneziano and Edward Witten. Unfortunately, neither of these theories have been proven to date. There was a lot of confusion regarding gravity, including the concept of curvature of space-time. Its unification with the others, if it happens, will make a Theory of Everything (TOE).

THE ULTIMATE LAWS

Now that we have unified the forces in nature, are we done? But before anything, unification or multiplication, we must answer the question, what is force?

This is something fundamentally present in nature. We have our hat on our head due to force; we can write with chalk due to force; our trains, vehicles and all locomotives move due to force. Humans took more than a thousand years to understand the true nature of force. The great conqueror Alexander the Great's mentor, the great philosopher Aristotle was also bothered with this question. Finally, he concluded that force is something that is required to cause any motion. You apply force to move one heavy box from one end of the room to the other. It was very close to the classical understanding of force, except Aristotle overlooked the fact that we have to constantly apply force to overcome friction which opposes the motion. The Father of modern physics, the Italian engineer and astronomer, Galileo di Vincenzo Bonaiuti de' Galilei came to a crucial conclusion. He stated that if anything is in motion unless an opposing force is applied, it will remain in motion, without the requirement of any other force.

Galileo died in January 1642. It was just one year after his death, a premature baby was born. It was said, he was so small that he would fit into a giant beer mug of about a litre. His father died before his birth; he grew up to be an aggressive child. When he was three, his mother remarried and this boy disliked his stepfather intensely. So much so, that he started hating his mother for marrying this man. His anger led him to threaten his parents with burning down their house. He was removed from his school and went on to do farming,

Newton was a person with varied interests; he studied alchemy and biblical chronology. Of the estimated 10 million words in Newton's papers, about one million deal with alchemy! He was the warden and master of the Royal Mint for the last three decades of his life.

but he hated it! After some persuasion by his teacher Henry Stokes, he went back to school to complete his studies. From then on, his studies were destined to change the world. Sir Isaac Newton Junior was one of the world's greatest mathematicians and physicists, also an alchemist, philosopher, a great believer in God, and lastly, the developer of classical economic ideas. One of his most extraordinary insights was on 'force'. He appreciated Galilean inertia for moving bodies, generalizing it for objects at rest as well. Things in motion remain in motion without deviation or change in velocity, and things at rest will remain at rest until an external force is applied upon them. This became his First Law of Motion.

Before moving on to his Second Law of Motion, let us discuss a concept extensively used in physics. Imagine you have a glass window, you have a light sponge ball and a heavy iron ball. You throw both at the same speed at the window. Which one do you expect to break the window? The heavy iron one, of course. So, a larger mass has a greater capacity to break. Now imagine two plastic balls weighing less than the iron ball but more than the sponge ball. You throw one lightly and the other at great speed. Now which, do you expect to break the window? The one that was thrown with greater speed. So the higher the speed, the larger is its ability to break. A new quantity, for simplicity, let us call it breaking capacity, can be conceptualized. It was named momentum, which is simply mass times velocity. Now imagine a body is travelling in open space (with no force acting on it). If its speed has to be increased, more push would be required. More push will also be required if the mass is more. Newton realized that force experienced by a body is essentially the change in its momentum. It was, however, not a complete answer to the question.

Imagine a force speeds up a body in one minute; is the same force required to speed it up to the same extent in one year? The answer is a big NO. However, the same energy is required. Newton concluded, for the first time, quantifying force and establishing one of the fundamental concepts of physics: force is the rate of change of momentum over time. This was the Second Law of Motion.

Newton took it a step further. He assumed that the mass of the body remains constant when a force acts on it, and force only increases its velocity. The rate of change of velocity with time is acceleration. Assuming this, he defined force as mass times acceleration. While this is true for most situations that we encounter in our everyday life, this was indeed an error at very high speeds. Newton did not consider the possibility of change of mass at very high velocities, as we know from the special theory of relativity, today.

Lastly, he showed that every force generates an equal and opposite reaction. Imagine a ball hitting a wall; it bounces back: why? Because the wall applied equal and opposite force upon the ball. This is Newton's last law that deals with force, the Third Law of Motion. Having established the nature of force, let us eventually see where and how it occurs in nature.

DELVING DEEPER

Now we know what force is, and we also know the fundamental forces of nature. Let us explore some of their derivatives in classical mechanics. What misled Aristotle could now be explained with Newton's third law. Friction is a resistive force, and it is because of this force, to keep a body in motion on the ground, there is a need to continuously apply a force on the body. To explain why friction, we need to zoom in, zoom to the point of contact between the surface of the body and the surface of the ground over which it is moving. When we push or pull, fine ridges at the surface of the body (as nothing can be absolutely smooth) rubs against the ridges on the surface of the ground. The force on the body is towards its motion, while the reaction force from the ridges on the ground acts in the opposite direction. A minimum amount of force would be required to overcome this resistive force and move the body. But if it is only the ridges that oppose the motion, then why do we need so much force to move a ship on the oceans; oceans do not have ridges! This question was solved with the discovery of viscosity, the resistive force in fluids.

What is this viscous force? Let us take a small wooden wedge and pour a few drops of alcohol and a few drops of honey at its top edge, ensuring that they do not mix. It is a no-brainer; we instinctively know who will win the race. The drops of alcohol will roll down the wedge much faster than the drops of honey. Again, it was Newton who explained why it happens. He showed that the fluids flow as if in infinitesimally small layers, with every layer above experiencing an opposing force from the layer below it, just like the force experienced by a solid body while moving on a surface. This results in the top layer being the fastest and the bottom-most layer, slowest. Force on a single layer was found to be proportional to the ratio of the velocity of the layer immediately above it and the velocity of the layer immediately below it. Force was found to be also proportional to the lower and upper layer's combined surface area. So, why did honey (as expected) lose the race? It is a fundamental property of a fluid. The force acting on a fluid layer is proportional to the factors discussed; to make it equal, a constant was introduced, which was given the name coefficient of viscosity. The coefficient of viscosity is unique to each fluid. Honey has a greater coefficient of viscosity and hence experiences more force, while alcohol having a lesser coefficient of viscosity experiences less force. However, one should not be misled into thinking that more force should give rise to more acceleration, as in the case of the example with the ball. As this is a viscous force, a drag force, acting opposite to the direction of motion of the liquid, it would tend to slow it down. If our ship was to sail on a hypothetical ocean of honey, its engine would perhaps require a large nuclear reactor to generate enough energy to overcome the viscous force!

BELOW WHAT WE "SEA"

How does the ship sail? To sail, the ship must first float. A ship of 200,000 tons, or even larger, how does it keep afloat? More than 2000 years ago, Archimedes of Syracuse, one of the greatest mathematicians and polymaths, encountered this question. Much before modern calculus which allows one to compute the surface area and volume of shapes and solids, was invented, Archimedes calculated the area of a circle, volume and surface area of a sphere also the area of an ellipse. He derived a fairly accurate value of pi (π), the mathematical constant which is ubiquitous in mathematics and physics. In Euclidean geometry, π is defined as the ratio of the circumference of a circle to its diameter.

Everybody then expected Archimedes to solve the mystery of the floating ship. We must remember that back then, most of the modern theories that govern the world were not known and unravelling the mystery was a big challenge. It occurred to him that "water, whether of the sea or a river or in your glass, is incompressible. Take a bucket full of water, for instance, and put a solid metal body in it. The body settles at the bottom of the bucket, water being incompressible rises by a height in the bucket. Archimedes realized that an upward force acts on the body in a vertical direction, tending to push it up. This upward force on the body is equal to the weight of the water displaced. But why then does the body not experience a force vertically in the downward direction? It does, but the lower face being at a greater height from the surface of the water experiences greater pressure than that experienced by the upper face, being closer to the surface of the water.

Archimedes came up with a groundbreaking idea. He said, why not make a hollow metal box? When the box is hollow, it still displaces an equal volume of water as before, but this time there is a significantly less downward force as the box has lesser mass (compared to the solid box) and, therefore, lesser weight. It became clear to him that the greater the volume of the body, the greater is the volume of water displaced, and greater is the up-thrust (upward force) and lesser mass and, therefore, lesser the downward force. This led to the birth of the concept of density, a new and very important concept in physics. Density can be quantified as the mass of an object per unit volume. So objects with lesser density float in liquids with greater density. Though this seems very obvious today, because we have been hearing it since childhood, it was a miracle discovery back then!

We come back to our ship problem. It is like a giant balloon, with a huge part empty, therefore even if its mass is 200,000 tons, the density is small, much smaller than the density of the water of the ocean as its volume is huge. So much that just the hull of the ship displaces enough water, causing up-thrust, resulting in a substantial part of the ship being above the water! And in fact, such a ship would be able to carry many sailors and huge quantities of cargo, as all of these together would only push the ship down into the water by a small amount compared to its height! And that is how all the maritime trade in the world is riding on an understanding developed by Archimedes more than 2000 years back!

Archimedes was the inventor of the mechanical devices that protected Syracuse during the seize. These included the Claw of Archimedes, a huge crane operated hook that lifted enemy ships and dropped them onto the sea. He also installed a giant parabolic mirror that concentrated the sun's rays onto the sails of Roman ships, burning them down.

The life of such a genius came to an end due to the arrogance and idiocy of a Roman soldier! The island of Sicily had an impregnable defence designed by Archimedes. During the siege of Syracuse (213 – 212

BC), Roman general Marcus Claudius Marcellus had given particular order not to harm Archimedes. But the Roman soldier, when he encountered Archimedes, commanded him to meet General Marcellus. The soldier did not recognize Archimedes. Archimedes, being preoccupied with a mathematical problem, declined to obey the soldier's command. Angered by this, the soldier killed Archimedes, who was 78 years of age then, with his sword — the end of a life... the beginning of a civilization. General Marcellus was angered and deeply disappointed with his death.

BLAISE, BERNOULLI AND BOEING

We now know why a 200,000 tons ship floats; the Archimedes Principle explains that. However, why does a 40-ton aeroplane made of metal fly? Though the weight of an average aeroplane is significantly less than an average ship, the concern is, it has to remain afloat in the far lighter air this time and not in the water! Water is almost 830 times denser than air. Another challenge for man!

Physicists came up with another observation, two things with the same weight, but one with a flat surface and another with needle end, both resting on your palm behave differently. The one with the needle end pokes you; it may even penetrate your skin, while nothing of this sort happens with the flat one. It was evident that a new quantity was required to explain this observation. It was defined as force per unit area, that is, pressure. A needle having a much smaller area of contact applies a much higher pressure on the palm, as the denominator, namely, the area of contact, is smaller. But wait, Newton's Third Law postulates that an equal force will act in the opposite direction. Since the palm will be having the same surface area of contact, it will apply equal pressure on the needle. So why is it that the palm does not penetrate the needle? The reason is that the hand is not a rigid solid. If instead of the palm, there would have been a metal sheet, the needle could have bent due to the opposing pressure.

The concept of pressure was taken a bit farther by one of the inventors of probability. French mathematician, physicist and catholic theologian, Blaise Pascal was studying fluid pressure, which led to his discovery of what is known as Pascal's law. Pascal's law states that pressure applied at any

Blaise Pascal was a child prodigy. He corresponded with Pierre de Fermat on gambling problems. This gave birth to the mathematical theory of probability strongly influencing the development of economics and social science.

point in a contained liquid is transmitted equally in every direction and on the walls of the container. The hydraulic lift was a device that makes use of this principle. This consists of a U-shaped vessel filled with water or some fluid such as oil. The two arms of the vessel are cylindrical with different diameters. Each arm of the U-shaped container is fitted with a closely fitting piston. A downward force is applied on the piston in the narrower arm, and the load is placed on the piston on the broader arm. The pressure applied on the narrower piston, that is, the force per unit area (of the piston) applied on it, is transmitted through the fluid in the tube to the wider piston in the other arm. As the cross-section of the wider piston is larger, the upward force is larger too on the wider piston (as the force is equal to the pressure multiplied by the cross-section area of the piston). The hydraulic piston, therefore, works as a force multiplier. We could, therefore, applying a little force, lift a heavy truck by using pistons of appropriate cross-sections!

In this example, what happens to the energy? Are we breaking the law of conservation of energy? We will have to consider the volume now; assuming water is incompressible, the volume of water pushed down in the narrower arm will move up the other wider arm and push the truck up. The arm with the smaller cross-section must be pushed down a greater distance to make up for the volume of fluid flowing into the arm with the wider cross-section. Energy use can be expressed as work done and work done is force times distance moved. The force required to lift the truck is much more, but the distance moved is less, while on the narrower arm, the force is less, but the distance moved is much more. Work done by the force acting on the piston in the narrower arm is equal to the work done by the fluid on the piston in the wider arm; thus, energy is conserved. Another miracle machine was invented where, by applying a small force, one could lift a huge load. All hydraulic presses use this principle. We encounter devices using this principle in many everyday situations; in factories to lift heavy loads and in automobiles in the form of hydraulic brakes.

Daniel Bernoulli's book Hydrodynamica published in 1738, announced another breakthrough. It was one of the finest applications of the law of conservation of energy. Applying the law, he showed that when a fluid flows through a pipe which

Daniel Bernoulli came from a family which boasted of at least eight notable scientists. As if that was not enough, their descendants were related to the Curie family, which had at least six notable scientists.

has varying cross-sections along its length, energy will be conserved at each point in the pipe with different heights and cross-sections. Bernoulli concluded that an increase in fluid velocity would result in a decrease in fluid pressure from the application of conservation of energy.

Now let us examine the overall structure of an aeroplane. It is so constructed that while flying, the wind must travel a shorter distance on the lower surface of the wings and a longer distance on the upper surface. Taking into account that an aeroplane while flying does not create any vacuum even temporarily, the wind taking the longer route has to travel faster than the shorter one to reach at the same time and not create a vacuum. From Bernoulli's principle, we know there will therefore be less pressure on the upper surface and higher pressure on the lower side; thus creating an upward thrust. Suppose the aeroplane reaches a very high speed, the higher pressure on the lower surface and the lower pressure on the upper surface of the wings will generate enough pressure difference, which will be enough to lift the aeroplane against gravity.

Though we have established the primary concept in lifting the aeroplane, we need such a massive body to run at a speed of 300 km/hr while taking off. After taking off, at higher altitudes with lower pressure and density of the air, it maintains a speed of about 1000 km/hr. This massive forward force is called thrust, generated either by the propellers on lighter planes or jet engines on heavier and faster ones. Thus, though Bernoulli laid

The plane Lockheed SR-71 Blackbird can fly 3500 km/hr, making it one of the fastest planes at almost three times the speed of sound in air.

the foundation for a flying machine, it took centuries for engineers to finally build the plane. Some of the pioneers who made significant strides to enable people to fly are the Wright brothers, Orville and Wilber, and others like Emma Todd, Victor Tatin, George Cayley, Frank Whittle, Hans Von Ohain, Albert Fono and many more.

But we also need to answer what helped the aeroplane to reach that speed. The meticulously engineered engine compresses, heats up and forces out air from a small orifice at the rare of the aeroplane. It is Newton again at work;

The Boeing 747, also known as the Dreamlifter has served as one of the largest passenger and cargo aircrafts. However, in terms of most metrics, the Antonov An-225 Mriya is the heaviest aircraft ever built with a maximum takeoff weight of 710 tons. It has the longest wingspan of 290 feet and six engines. Only one of this monster aircraft has been built.

his third law talks about the reaction force that pushes the aeroplane forward when the heated air is forced out of the orifice. While moving forward, the aeroplane experiences a drag force caused by an enormous number of collisions with the air particles through which it flies. This drag force has to be minimized; thus, aeroplanes are made of special design, in other words, aerodynamic designs, to minimize these collisions.

FURTHER ON FLUIDS

Are we done we fluids? Fortunately, or unfortunately, no. In fact, it is just the beginning. We began with Archimedes Principle, Pascal's Law, Newton's Law of Viscosity and Bernoulli's Principle and built a castle. Imagine you can track and trace any air molecule that passes by you; what do you expect its motion to be like? Fluids became a central field of study, as nature is predominantly in the fluid state, the air in our atmosphere to water which covers 70% of the surface of the Earth, 60% of the weight of your body, to the coffee you drink.

Nevertheless, how much of their character do we know? What shapes of fan blades would make the most efficient ceiling fan? What shape of a boat would cause the least drag making it faster and more efficient? When should one warn citizens of an impending storm? When should a dam be opened to prevent a flood? What will be a doctor's prediction on blood leak from an artery? How will the sloshing of liquid fuels in cars, trucks, planes, rockets be and how would they affect their motion?

One of the greatest minds in history attempted to bring all of these factors into a mathematical equation that will predict the flow of fluids. We will come across this person, again and again, as we study real-life problems and eventually admire his multifaceted genius. Born on 15th April 1707 in Basel,

> *Euler became blind in both his eyes. Yet he worked in almost all areas of mathematics and was prolific. After he became blind, he became even more productive; in 1775, he produced almost one mathematical paper every week!*

Switzerland, Leonhard Euler came up with the first fluid equation. However, his assumption that inviscid (non-viscous) fluid would obey his equation did not work out. It was a teacher of Laplace, the discoverer of the solution to the one-dimensional wave equation, Jean le Rond d'Alembert, who proved that Euler's equation predicts a zero drag force which was in contradiction to observation. Today, this is known as d'Alembert's paradox.

This paradox was resolved, and the final brick in the castle was placed by the French engineer Claude-Louis Navier and Anglo-Irish physicist Sir George Gabriel Stokes. Their equation, popularly known as the Navier-Stokes equation, answers all the questions about the fluid flow, hence its iconic status in the area of fluid flow. However, the beauty of the Navier-Stokes equation lies in its ability to describe all the fluid motions, taking into account viscosity; flows such as the rings of smoke blown out by a cigarette smoker or the rainwater overtopping the drains in a town during heavy rains!

What does the Navier-Stokes equation say? It is a partial differential equation that, with given boundary conditions, would predict a map across the body of fluid under study. This map has time-variant vector fields (velocity) at every point of the fluid; that is, you consider any point in the fluid at any point in time, and you will be able to calculate at what speed and in what direction the fluid is flowing through that point. The equation being time-variant will take into account changes with time. Navier and Stokes opened the gates for solutions to more than a million engineering, medical, physical and astronomical problems. But was it enough for the scientists? Did we unravel the mysteries of fluids, or did we expand the universe of mysteries? It is a deep question.

IS THE NAVIER-STOKES MAP 2-D?

Navier-Stokes equation successfully described two-dimensional flow. It can also be used to simulate three-dimensional fluid flow. But unlike the two-dimensional solutions, the three-dimensional solution does not come with the existence and smoothness proof. These became fundamental problems, not only to physics but also to mathematics.

By existence, it is meant, in a simple way, that the pattern the equation gives for 3-D flow at a given time has at least one value and direction of speed of the fluid contained in the volume under consideration.

By smoothness, it is meant that there is no mathematical singularity in the value and direction, or let us say that under the same conditions the solution of the Navier-Stokes equation will always give a definite answer for the value and direction. This would mean that the solution to the equation at different times, using the same process and conditions, shall always be the same.

The proof of existence and smoothness in the Navier-Stokes solution in 3D is one of the most spectacular problems in mathematics and physics. In the year

2000, Clay Mathematics Institute included it in one of the seven Millennium Problems. Six of these problems, including the general solution to the Navier-Stokes, have eluded solutions so far. Solving any of these will win you a million dollars; you may like to have a go.

Another great mystery became even more mysterious with the Navier-Stokes equation. Through computation, the equation may predict turbulent flow in two dimensions, but the accuracy of the prediction in both dimensions remains unknown. Thus, a mathematical and intuitive interpretation of turbulent flows still remains the greatest unsolved problem in classical mechanics.

A dimensionless number depending upon inertial and viscous forces called Reynolds number, which has a significant role to play in fluid motion, was discovered by Sir George Stokes and Irish engineer and physicist, Osborne Reynolds. This magical number is a property of every fluid. If the Reynolds number exceeds a certain value and keeps on increasing, something strange, incomprehensible in short, a magical thing happens! A streamlined or laminar flow changes into a turbulent flow! This number has huge significance for ship designers who could now make a small scale model of ships and utilize the data obtained from using Reynolds turbulence principles to predict the behaviour of the full-scale ships. Richard Feynman, Werner Heisenberg, along with many others, were deeply interested in turbulent flow.

A person with severe anxiety disorder, schizophrenia, bipolar depression, epileptic disorder, perhaps more than these, who failed in all his works, ultimately at the age of 28 held the brush and made over 2000 paintings, it was Vincent Wilhelm Van Gogh. A year before he committed suicide at the age of 37, seated in a lunatic asylum, created a painting. This painting, titled

> *Loving Vincent is an experimental biographical drama film about the life of Vincent Van Gogh. Each of the film's 65000 frames is an oil painting on canvas created by 125 artists from 20 countries using the same techniques as Van Gogh! The film won several awards and received good viwer reviews.*

Starry Night, captures a very important and little understood physical phenomenon that was observed, arguably the most famous painting by one of the greatest modern painters! Why this particular painting in the middle of a severe turbulent crisis? Van Gogh who severed his left ear during a mild seizure-like attack was admitted to a mental asylum. There, seated on the ground, he created this painting! Centuries later, it caught the attention of leading scientists when they discovered stars moving in eddies, as depicted in the painting. Eddies are one of the chaotic outputs of turbulent flow, where the fluid flows in a circular motion. Eddies vary in size. But what made Van Gogh paint those stars and sky in eddies was indeed a matter of great surprise for the scientists.

Soviet mathematician, Andrey Nikolaevich Kolmogorov achieved a significant breakthrough in the field of turbulence in the year 1941. His understanding and prediction of energy distribution, varying through the size of eddies, was indeed the work of a genius.

Talking about the turbulence and its existence in celestial movement, where else do we find it? The answer is, everywhere there is fluid, wind, smoke, tea, milk, river, and anywhere you look. Even where you cannot see, inside your body, inside the arteries, there is turbulence. Even the blood which flows through every vein and artery in our body can exhibit turbulence!

WHEN TWO ENIGMAS MEET!

Blood flowing through the veins and arteries shows mysterious turbulence. It would be good to know what blood is. Though more than a million people throughout history were slain, and proverbially rivers of blood flowed, it was not until the English physician William Harvey discovered the systemic circulation of blood that people learnt its true nature. Systemic circulation means blood being supplied to all body tissues, like the brain, muscles, kidneys, liver and the heart itself! The heart is the pump that is responsible for forcing blood all across the body, it is the powerhouse of circulation. Blood, more specifically mammalian blood, appears red due to large numbers of a particular type of cell that contains a protein molecule, haemoglobin giving it its colour and its principal task.

Blood has always been a matter of great interest through the ages and finds mention in literature in all cultures. We have heard of vampires like Dracula, a character from Bram Stoker's novel, which was said to suck blood from its victims. In European folklore, vampires were once upon a time thought to be real. Vlad Dracula, after whom Stoker named his character, supposedly drank blood! The ruler of Romania, Vlad Dracula, made bloodshed a daily ritual in his kingdom. Many years after him, also in Romania, the noblewoman Elizabeth Bathory was suspected of bathing in young people's blood to maintain her youth! Their cruelty can never be justified; however, we can safely rule them out at least from drinking blood! Indeed, if one drinks quite a lot of blood from someone, he would require immediate consultation with a doctor. Not the victim, but the drinker, as blood, can cause iron overdose problems when ingested in large quantities. This is because every haemoglobin molecule that carries oxygen and carbon dioxide in the blood contains iron. Surprisingly, even as late as 1959, the structure and functions of this molecule were not fully understood; we now understand it quite well.

British surgeon and physiologist, William Hewson is referred to be "the father of haematology", the study of blood. He described blood coagulation, an essential safety mechanism for the human body and isolated fibrin, a key protein present in blood. His discovery indicated that there are in blood many mysteries beyond their colour red.

RUNNING ON RED

Before probing further, let us explore why blood appears so red. There are close to 5 million red blood cells, or erythrocytes, in one millilitre (one-thousandth of a litre) of blood. The erythrocyte is a disc-shaped cell, and each cell contains approximately 270 million haemoglobin molecules, the active

ingredient responsible for its characteristic red colour. Every haemoglobin molecule with four heam groups can bind with four molecules of oxygen. You are carrying almost 5 litres of blood in you.

Nevertheless, one may ask, why so much, why more than a gallon? We need the energy to walk, to run, to talk, in fact, for every conceivable activity. Moreover, this energy comes from the food we take, the fats and carbohydrates. A car needs energy to run. In the internal combustion engine of your car, the mixture of air and gasoline fuel is ignited. The oxygen in the air breaks down the carbon-hydrogen bonds in the fuel in an exothermic reaction called combustion. The energy released leads to the heating and therefore expansion of the gases, which push the pistons to cause motion. Similarly, every individual cell, which is alive and active in you (not the dead ones such as the hair), requires oxygen for its energy. Each cell has its own internal combustion engine, the mitochondria. Furthermore, the required oxygen is transported to the cell by the haemoglobin in the red blood cells of the blood. Of course, some cells do not receive oxygen carried by haemoglobin; such as the cells of the cornea, the outermost layer of the eye, which can absorb oxygen directly from the air! Besides such exceptions, God has made a tremendously complicated architecture of tiny vessels called capillaries which reach out to every second cell inside you. Though the geometry of distribution of the capillaries may differ from person to person, the macroscopic arteries – vessels taking blood away from the heart and veins – vessels carrying blood towards it, are nearly the same in all humans with normal circulation. Thus, there has to be enough blood in the body to be able to supply all cells with oxygen so that the mitochondria can produce the energy.

Thankfully, God even planned to transport the carbon

dioxide, one of the end products of the energy-releasing respiration, out of our bodies. The carbon dioxide is partly dissolved in the blood, partly converted to bicarbonate and partly absorbed by haemoglobin and released in the lungs and expelled from the body during exhalation. If this did not happen, then the human body would have had to store all the oxygen molecules, whether reacted or unreacted. The source of the two oxygen atoms in carbon dioxide is understandable; where did the carbon come from? One of the reasons you lose weight by not lazing around is because of the increased rate of respiration, which is the burning of carbohydrates and fat and the eventual expulsion of more and more carbon from the body. All fats, polysaccharides, sugars finally break down to glucose, a molecule with six oxygen, six carbon and twelve hydrogen atoms. During respiration, the oxygen absorbed by haemoglobin reacts with glucose, breaking the bonds of the big molecule. A glucose molecule consists of six atoms of carbon, twelve atoms of hydrogen and six atoms of oxygen. The six carbon atoms from it combines with six oxygen molecules brought by haemoglobin and form six molecules of carbon dioxide, which makes its way out of the body. Carbon came from glucose, but what happened to the hydrogen in the glucose? These hydrogen atoms, paired up with the six oxygen atoms, forming water. This water is used in the body or maybe expelled out. Of course, the actual reaction is not as simple as that; respiration consists of dozens of intermediate reactions before carbon dioxide is formed.

BEYOND THE RED

The blood flows at an average speed of about 6 km/hr. The heart pumps about 5 litres of blood per minute. Both the kidneys receive about 1.25 litres of blood in a minute, filter this blood, and the waste products are excreted out with excess water. The heart pushes about the same amount of blood into the two lungs, where the exchange of gases takes place. Moreover, the heart sends about 0.75 liters of blood to the brain on an average per minute. Is it, then, the heart that determines all the movement of blood across the body?

Before answering, let us try to find that part of blood hidden among the red cells. Your hand has suffered a cut, it starts bleeding, you do not have any aid at that instance, and you fear all your blood will drain out of the body. However, that does not happen. In nature, there are sharp things such as thorns on cacti, rocks with sharp edges, etc., that cut through the skin of animals. They do not use bandages, so how does the blood leakage stop? The blood perhaps is acutely "aware" of the possibility of draining out completely, so it carries something called platelets that gets triggered when they encounter an unnatural roughness like a cut in a blood vessel. The platelets begin to create their own bandage there, with one platelet summoning another through chemical signalling. The platelets go through several reaction steps forming a net-like film called fibrin. This is the natural bandage for the cut blood vessels, and the bleeding stops.

Platelets are efficient, but if the cut is large or very deep, they fail to create nature's bandage to stop bleeding. That can cause serious problems. Sometimes platelets get confused and start the whole process of assembling and bandaging, where it is not needed, causing unnecessary blockage to the path of blood flow; these are called blood clots or thrombus. In these circumstances, these platelets, as before, call more and more platelets, ultimately causing a significant amount of blockage, restricting the flow of blood to the tissue supplied by the vessels. With the stoppage of blood flow, the supply of food and oxygen to the designated tissue also stops; cells being starved of food and energy die. Even if not consistently effective, the platelets are a remarkable self-protecting mechanism of the body. But platelets cannot operate instantaneously, the summoning of platelets and chemical reactions take time. If the surface causing the cut is soiled or infected, it is not free of microorganisms. In the time the platelets take to create the natural bandage, harmful microbes get enough time to make their way into the body with adequate nutrition and room for multiplication. Even with the smallest of cuts, we would have been destined to be the victims of our microscopic enemies if God did not provide us with some armed companions of platelets in our blood!

TRAINING OUR PROTECTORS

Before visiting the other components of blood, we need to know about the work done by the Father of microbiology, the Father of immunology and the Father of bacteriology. Their works are so closely associated that their fields overlap.

The first, Antonie Philips Van Leeuwenhoek, a Dutch cloth merchant, was exploring ways to have a better view of the warp and weft in pieces of textiles. He took an interest in lens making and finally ended up inventing the microscope. With his invention of the microscope, he surely had a better view of many things than ever before. He took particular interest in those living creatures our naked eyes had never seen! With his new invention, he studied blood, muscles and discovered many other biological things, even the listing of which would take a whole book!

The second, English physician Edward Jenner, noticed that milkmaids who had earlier contracted cowpox were mysteriously safe from smallpox even during a smallpox epidemic. He believed that it

The smallpox virus is a close relation of the cowpox virus. 'Cow' is 'vacca' in Latin, and cowpox is 'vaccinia'; hence the term vaccination.

was something to do with their profession, maybe something in the cowpox that protected

them. Much before this observation of Jenner, people chose to deploy surviving plague patients to treat those infected, knowing they would not get infected again! It was Jenner who connected these facts and conjectured that our body perhaps has a memory, a memory that protects us from contracting the disease. Closely analyzing the milkmaid conundrum, he concluded that the milkmaids infected with cowpox must have developed something in their body which protected them against, smallpox which was similar to cowpox. This observation gave birth to a revolutionary idea. He administered pus from cowpox blister from an infected milkmaid to a child. Though the child contracted cowpox, once his fever subsided, he did not contract smallpox even when exposed to the disease! It was the world's first vaccination.

A way had been found, but still, many obstacles remained. Ultimately, it was a professor of physics, chemistry and, geology who came up with a breakthrough in biology! Louis Pasteur, an average student in his early years, interested in fishing and sketching, is regarded as one of the fathers of the germ theory of diseases. His works led to the development of vaccines for rabies and anthrax, saving millions of lives. His research led to the development of a technique of treating wine and milk to stop bacterial contamination, the process now being called pasteurization. Pasteur made major contributions to chemistry as well, particularly in understanding optical isomerism. It is beyond the scope of this chapter to list his enormous contributions to different areas of science.

Having learnt a bit about Pasteur, let us turn our attention to one of his students. In the spring of 1894, he arrived in Calcutta carrying with him another vaccine to prevent one of the most life-taking diseases, cholera. However, he faced resistance from the British; for one, he was not a doctor and secondly, he was an East European. Indians were hesitant, too, because they were scared. He was the Ukrainian

The first cholera pandemic occurred in Bengal in India and lasted from 1817 to 1824. The disease spread to South-East Asia, Middle-East, East Africa and Europe through trade routes. Hundreds of thousands of Indians and about 10000 British troops died in this pandemic in India.

zoologist and bacteriologist, Waldemar Mordechai Wolff Haffkine. To allay fears, he tested the vaccine on himself to prove its effectiveness. Quite soon, he was invited to Bombay during a plague outbreak to demonstrate another of his inventions, the plague vaccine. As before, he tested it upon himself first. Then he spread the vaccine among the people to save the world from these two life-threatening diseases of the time. For his exceptional contribution, Sir Joseph Lister named him "a saviour of humanity".

Sir Joseph Lister was a British surgeon and the pioneer of antiseptic surgery, who applied Pasteur's advances in microbiology in the practice of surgery. His work reduced risks for patients undergoing surgery and

> *Lister used carbolic acid for disinfection and it had its problems. His ideas on antiseptic use and transmission of infections were mocked at. Even The Lancet, the prestigious medical journal warned the medical fraternity against adopting Lister's ideas!*

distinguished him as the "father of modern surgery". Each of these great inventors contributed the pillars on which modern medicine rests today.

Though it may seem that the procedure adopted by Jenner can be applied in all diseases, it is not so. All vaccines cannot be made in the same manner; modern vaccines are prepared through dozens of reactions and processes in highly sanitized laboratories. It is a hugely painstaking work by teams of excellent researchers working in modern laboratories often over several years.

THE WARRIORS WITHIN

Before we make an acquaintance with our microscopic warriors, who were either triggered or trained through the efforts of the path-breaking scientists mentioned in the previous sections, it would be in order, to name a few more who made huge contributions in the field of immunology. Robert Koch, Theodore Schwann, Paul Ehrlich, Richard Pfeiffer, Karl Landsteiner, Rosalyn Sussman Yalow, James Gowans and many others advanced our knowledge in immunology and made us face the challenges of newer and newer diseases.

Let us zoom in to the warriors in our immune system, collectively called the white blood cells (WBC) or leukocytes. The white blood cells are classified into three broad categories *viz.*, granulocytes, monocytes, and lymphocytes.

Granulocytes can be further classified into neutrophils, basophils and eosinophils. These mostly perform phagocytosis; that is, they devour intruders such as bacteria and eliminate cell debris. However, they are also responsible for inflammation and allergic reactions in the body during their fight with intruders.

The big hungry cell, macrophage, a monocyte, is a major phagocytic cell. When the intruders overwhelm the granulocytes, exceeding their killing capacities, monocytes appear on the battlefield. The dendritic monocytes are cells with dendrites, that travel among the tissues reaching the lymph nodes and calling on the lymphocytes to join the battle.

The T-lymphocytes, which are produced in the bone marrow and mature in the thymus gland receive messages from the dendritic cells as to the nature of the invaders that is whether they are like fungi, bacteria or viruses. Once the T cells know about the invaders, some of them start releasing chemicals called cytokines, which triggers other leucocytes, including B-lymphocytes which remain in the bone marrow. B-cells, on receiving the signals, get activated by multiple manifolds and start making a Y-shaped protein and its derivatives called antibodies. These antibodies storm the battlefield and attach themselves to the invader cells, either disabling them or enabling the phagocytes to engulf them. Another type of T-cells, the natural killer type, release their own molecules to kill the invaders. Finally, some T-cells work as generals and guide others so that they do not kill the body's own cells. This is the immune system we are born with; the innate immune system.

The reason why surviving plague patients were deployed to treat people suffering from the plague, the reason why people who have had chicken pox once do not suffer from it again, and the reason that vaccines protect us from diseases is because of something called acquired immunity. Nevertheless, we are not born with acquired immunity. The last type of lymphocytes is the memory T and B cells which remember the nature of the invaders. Once a person is infected with some disease, these memory cells keep a note of the characteristics of the enemies. If these cells happen to intrude into the body again, these memory cells recognize them instantly and kill them, even before we notice them.

This is a very simplified model of our immune system, an army at the microscopic level with miraculous powers which protect us from diseases Every day that we are alive.

BACKUP FOR THE REGIMENT

Alexander Fleming discovered that lysozyme, a killing agent found in mucous, tears *etc.*, can effectively kill bacteria. It is, therefore, a part of the body's immune system itself. But later, on returning from a holiday, Fleming discovered that one of the bacterial cultures he had kept before leaving for his holiday was infected by a fungus. The fungus was identified to be of genus *Penicillium*; there exist about 300 species in this genus. The species in Fleming's culture was identified as *notatum*, now identified as *rubens*. It had secreted something that killed the bacteria in his culture. He further noticed that this secretion killed bacteria which cause a variety of diseases like meningitis, scarlet fever, *etc.*

Born in Scotland in 1881, Sir Alexander Fleming, often referred to as the greatest Scott, had discovered the world's first bacteria killer or antibiotic! This is often regarded as man's greatest achievement against diseases.

Fleming showed the way to kill bacteria, revolutionizing medical science. Nevertheless, *Penicillium notatum* was not good enough to kill all species of disease-causing b a c t e r i a . E v e n t u a l l y , hundreds of antibiotics were discovered and are still being d i s c o v e r e d .

French bacteriologist Jean Paul Vuillemin introduced the term 'antibiosis', meaning "against life", in 1877. Selman Waksman and his collaborators first used the term antibiotic in journal articles in 1942 to describe any substance produced by microorganisms that is hostile to the growth of other microorganisms.

Along with antibiotics, antifungals to fight diseases caused by fungi and antivirals to fight diseases caused by the virus were discovered. Even if the natural immune system fails, we have a major backup, a gift from science.

Have we conquered everything? Unfortunately, not yet. We cannot forget our old friend Darwin and his natural selection. Bacteria are living beings and, like any other living thing, they will evolve. Like us, they are winning over natural hurdles over many generations that have evolved enough to fight present obstacles. Similarly, some surviving microbes, selected to be fit enough to survive the antibiotic, will multiply and pass their traits to their newer generations. They will eventually overpower the currently available antibiotics and be more potent and deadlier as they multiply manifolds within no time. That is why patients are called upon to complete antibiotic doses to kill the targeted bacteria in their systems. Professional medical bodies advise patients not to self-prescribe antibiotics and also not to take incomplete doses. All of these can give the germs a chance to evolve into more menacing variants. This was not really an end to the fight against diseases but in fact, a beginning. As bacteria evolve, scientists will continue to search for newer and more potent antibiotics. Our immune systems are living too; God will re-engineer them to let us survive in this world.

THE INDEFATIGABLE PORTER

The nicotine inhaled (while smoking a cigarette), the syrup drunk, the dessert eaten, the insulin injected into our vein all rely on our natural transport system, the blood. Blood carries them to their desired organs, like excess glucose to the liver, nitrogenous wastes to kidneys and so on. There are many functions of blood, all cannot be discussed in the space of this book, but this one, transportation, is a fundamental function of blood and merits a discussion. Blood transports hormones. Hormones are special chemicals produced by endocrine glands and released directly into the bloodstream. Though there are other hormones as well, endocrine hormones have a vital impact on the physiological and psychological development of a person. The concentration of hormones in the blood has to be at an optimum level; a minor imbalance can cause hormonal diseases, some of which were studied and identified by the English physician and scientist, Thomas Addison and American neurosurgeon and pathologist, Harvey Williams Cushing.

Some hormones impact the working of other endocrine glands and, therefore, the production of other hormones. An example is, the thyroid-stimulating hormone (TSH), produced by the pituitary gland, situated at the base of the brain behind the nose bridge, it sends messages to the butterfly-shaped two-lobed thyroid gland located at either side of the windpipe, regulating its secretion of thyroxine that has a vast influence on nearly every cell of the body. These glands, being interlinked functionally, form a system called the endocrine system.

The endocrine system relies entirely on blood for the transport of its produce, but how does the blood know of the destinations of hormones released by specific glands? Let us consider the example of the thyroid-stimulating hormone released by the pituitary gland in a niche within the brain, which is required at the thyroid gland and not at the kidneys. Since blood is circulating all through the body, how does it know the destination of the TSH? In reality, the blood "does not know" and the endocrine glands "know" that the blood "does not know". It works like this; the pituitary gland secretes more thyroid-stimulating hormone than required by the thyroid, recognizing, that the blood will take some of it to all parts of the body where it would not be required. However, it works only in the thyroid gland: why? Imagine for a moment that the pituitary had sent a box with a letter, and only the thyroid has the key to open it and read the letter. The thyroid is thus able to open the box and read the instructions that the pituitary has sent, and it reads "produce more thyroxine". The other cells in the body would also have received the box, but as they do not have the key, they would not be able to open the box at all. Actually, this analogy is now central to this mechanism called the lock and key model. The key and the lock are different chemicals, which, when they meet, get attached to each other and react in a particular way, sending some specific message somewhere. Our body has an elaborate "postal system" which ensures that messages sent by one set of tissues are acted on by the addressees and no one else!

Let us explore our analogy a little more deeply. Let us believe now that the many locks are fitted in the body, can there be any skeleton key to open all these locks? Unfortunately, the answer is "yes", and unfortunately, the popular drug nicotine (present in tobacco) is one such skeleton key! If smokers quit smoking, they could avoid the damage to their bodies from the smoke. Although the brain is the desired location for nicotine, blood will take it everywhere, to the heart muscles and arteries and also to the gastro-intestinal tract, all of which have locks that can be opened by this key! Though an overall pleasant feeling is attained as nicotine reaches the brain, the other locations suffer damage due to nicotine.

Natural hormones created in the body have the ideal and efficient lock and key system. Not only humans but every animal and every plant is influenced by their hormones, from their structures to their vital functions. We now turn our attention to plant hormones as that would give us an important insight into the role they play in plants and how plants differ from animals.

FROM SEED TO TREE

From Lamarck's second law, we know a species will transfer its traits to its offspring. We know a Royal Bengal tiger's cub will usually be striped, and a Black Bengal goat's kid will usually be black. Under normal conditions, the next generation of humans will stand erect on two legs and will be endowed with two hands attached to the shoulders and have two eyes, two ears and a nose on the face. We are not talking about physical deformities here, which are exceptions. But this feature does not hold good for trees. We can recognize the 'offspring' of a mango tree as a mango tree, but the young mango plant is not a replica of the parent mango plant. That is, the young plant will not have the same number of branches emanating from the same positions on the trunk or possessing the same number of leaves! To clarify further, a parent mango plant has 20 branches, while another plant that had grown from its seed has 17 branches with completely different orientations. The texture, leaves, fruiting, colour of the "offspring" plant will indicate both are mango trees; yet something has happened which does not make them identical!

Unlike in animals, plants do not make carbon copies of themselves; in plants, the structure is entirely determined by their hormones. In turn, hormones are determined by the nature or environment around them. Plants do not have a highly developed circulatory system that could carry the hormones around the different parts of the plant. Though tissues like xylem and phloem are present, xylem mostly carries water and nutrients from the soil to the branches and leaves, and phloem ordinarily carries food synthesized by the plant; hormones do not have a dedicated transport system. Hormones generally travel from a plant cell to another cell by means of diffusion – this mode of transport results in hormones being confined to particular regions of the plant body.

Plants exhibit tropic movements, which refers to change in the direction of growth to various stimuli; these are largely due to hormones. Examples of tropic movement are thigmotropism (or haptotropism) which refers to change in orientation of growth as a reaction to touch, geotropism (or gravitropism), which is the growth of a plant's organ as a reaction to gravity, phototropism or growth directionally in response to light. Let us try to understand how this works at the molecular level. We notice phototropism through the bending of a shoot of a potted plant placed near a window towards the window. Tips of plants and branches produce amazing light-sensitive chemicals called auxins. The most active auxin is indole-3-acetic acid, a plant hormone. This hormone promotes the growth and elongation of plant cells. However, how does the plant know which way to bend in the presence of light? The indole-3-acetic acid is actually repelled by light. When light is incident on one side of the apical tissue, this chemical diffuses to the side of the growing shoot away from light. As this auxin promotes cell growth and elongation, the cells on the side away from light grows and elongates. This unequal elongation, more on the side away from light, results in the plant bending towards the light. Our childhood magic of plants growing towards the light, even from a hole in a dark box, is now explained! Though the same hormone can act differently in other parts of the plant to promote other tropic movements, they essentially work by reacting to light's energy.

When the plant bends towards the light, it needs to grow as well. Gibberellins are hormones that help in cell elongation with cytokinins causing cell divisions leading to the growth of branches and leaves. These branches also bend towards light due to auxins.

There are many plant hormones, but why their collective effect makes the plant, with or without branches, grow towards light? Leaves want the light to carry out photosynthesis and prepare food. We all know that each leaf in a plant is a food factory, but given that plants do not locomote, talk, play, why do they require so many food factories in them?

FOOD FACTORIES

The chloroplast inside a plant cell, predicted by Russian biologist and botanist, Konstantin Sergeevich Mereschkowski and observed by German botanist and phytogeographer, Andreas Franz Wilhelm Schimper, closely resemble cyanobacteria, a type of bacteria that has a substance called chlorophyll. It was later conjectured that it was a bacteria, which invaded a plant cell between 1 to 2 billion years ago, but for some obscure reason, neither did it attack the plant cell nor did the plant's immune system attack it. These bacteria became an advantage, producing food for the host cell! Over time the cyanobacterium was assimilated and many of its genes were lost or transferred to the nucleus of the host, and it became the chloroplast! Another type of bacteria invaded both early plants and animal cells (or their common forefathers) and became an essential part of cells: the mitochondria. How is it that we know they are foreigners to our cells, and why are they residing peacefully, not attacking us, and why is our immune system not getting rid of them? This phenomenon is known as endosymbiosis, where a foreign cell lives within another cell with benefits for both!

They are perhaps one of the best and earliest examples of symbiotically living organisms! It is not that there is something wrong with them. Perhaps, it is our assumption, that if any foreigner invades us, we need to attack them to save ourselves is wrong! We have spoken of glucose being broken down with oxygen releasing energy which is essential to our life. This energy is temporarily stored in a chemical called adenosine triphosphate (ATP) which takes it to the desired site for use. Mitochondria

in the cell are the site where this chemical ATP is synthesized. Thus mitochondria keep us alive, and in turn, we offer them a home! How do we know that mitochondria are foreigners to our bodies? Apart from their structural and chemical similarity with bacteria, they have their own genetic materials. Its vital role is now understood from the contributions of scientists like Hans Krebs, Gustav Embden, Otto Meyerhof, Jakub Parnas, John Walker and many others.

Now we go back to our first concern, the food factories or chloroplasts. Chloroplasts are found only in plant cells. They have their own genetic material, indicating that, like mitochondria, they are also foreign to the plant cells. They help plants "cook" food, and in return, plants provide them with a home! With contributions from Jan Ingenhouz, Melvin Calvin, Rudolph Marcus, James Bassham, Andrew Benson, and many more, we are now able to know how the food is cooked inside a plant cell by a process called photosynthesis.

But why do plants need food when they do not need to run, exercise, and think like us? Why are they such philanthropists, preparing food not only for themselves but also for the whole world? Answering the first, they need food to grow, repair, and reproduce. Answering the second can be a little more taxing to the brain. It emanates from the primaeval urge of all living organisms to multiply and propagate. Plants produce nectar, a food especially loved by insects like bees and butterflies. They visit flowers for nectar, and while at it, they pick up pollens. When moving from one flower to another, they transport the pollen and help in fertilization. Fertilization produces seeds, and seeds need a lot of food to survive and germinate. Plants have to produce hundreds, even thousands of times the seeds they actually require to reproduce, as many of them are lost, many devoured by birds and other animals, and many do not get a conducive environment to germinate. In many cases, the seeds are covered by fleshy tissue, which we call fruit. Fruits are one of the primary foods. Plants produce fruits in their own interest; after all, animals, after consuming the fruits, reject the seeds which germinate, enabling plant proliferation. It is also possible that plants do not intend that animals consume the food prepared by plants, but animals eat them nevertheless. Thus, plants become the world's primary food supplier.

Where do plants find all the raw materials for cooking their food? The raw materials, carbon, hydrogen, nitrogen, oxygen, and other minor nutrients are everywhere in nature; in the soil, in the water, and in the air around us. These are the ingredients and spices which plants use in their kitchens, the chloroplasts to prepare the food! Furthermore, for the stove, they have the inexhaustible sun above!

We have many sources of energy in nature – coal and petroleum which are exhaustible and wind, sun, sea waves, which are non-exhaustible. But the real source of all energy is the sun. It is pretty amazing; an outer space stove is the source of all energy on Earth. What gives the sun all its energy?

THE FLICKER OF FUSION

It seems somewhat an obvious fact today, that the sun is a giant nuclear bomb! It was not so back in the sixties until the German-American nuclear physicist, Hans Albrecht Bethe and his team discovered nuclear fusion reaction in the sun. As in the Tsar Bomba, billions and billions of atoms of hydrogen fuse to form helium atoms, accompanied by the emission of tremendous energy.

If it is only hydrogen converting to helium, how is it that the sun continues to supply energy for so many years; after all, it is about 4.6 billion years old? Why does it not immediately burn out? The answers to these questions once again lie in one of our fundamental forces, gravity. The massive weight of the solar material and gases press on the core of the sun. This intense pressure compresses the material in the core, and it becomes so dense that

In the core of the sun, about 600 million tons of hydrogen fuses into helium every second, the nuclear fusion converting 4 million tons of matter into energy every second as a result. This energy, which can take between 10,000 and 170,000 years to escape the sun's core, is the source of its light and heat.

hydrogen molecules come close to each other at the atomic level. As a result, fusion reaction occurs in the core and not in the outer layers of the sun.

If hydrogen is the fuel and is exposed to extreme heat, and as we all know, hydrogen is a highly flammable gas, why does it not burn out? This was answered even before this question arose. Father of modern chemistry, French nobleman, Antoine-Laurent de Lavoisier,

> *Joseph Priestly invented carbonated water. Carbonated water is present in cold drinks, beer and soda, things we consume every day.*

defined what is burning or combustion. Fire in Indian and Greek philosophies have been considered a fundamental force of nature and was worshipped by them for ages. English chemist and philosopher, Joseph Priestly, a publisher of about 150 works, discovered oxygen. Priestley's works were translated from English to French by Marie Anne Lavoisier, the wife of Antoine Lavoisier, thus opening the doors for her husband to work with her and some other friends and prove that combustion is essentially a highly exothermic reaction in the presence of oxygen. The sun, extremely hot as it may be, cannot support combustion (of hydrogen) as it does not have sufficient quantities of oxygen. If we could supply enough oxygen to "burn down the sun", it would create a supermassive drop of water due to the hydrogen burning in the oxygen to form water!

Joseph Priestly's discoveries made way for Lavoisier, who unravelled some of the greatest mysteries of nature, identified and named oxygen and hydrogen, and helped construct the metric system and reform

> *Lavoisier was charged with adultering tobacco with water, though in reality, he devised a very rigorous system to prevent adulteration. After a year and a half of his execution, Lavoisier was entirely exonerated by the French government. Unfortunately, it was too late!*

chemical nomenclature. Unfortunately for science, Lavoisier's life was cut short on the 8th May 1994 at the guillotine, a device devised by another Joseph, Joseph Ignace Guillotine, during the French Revolution.

Famous Italian mathematician and astronomer, Joseph-Louis Lagrange lamenting Lavoisier's death at the guillotine, said, *"It took them only an instant to cut off this head, but France may not produce another like it in a century"*.

Coming back to our sun, the awesome heat and light produced in the nuclear fusions reactions in the core travels to the surface of the sun in a minimum of 10,000 years. On average, this light takes about 8 minutes 20 seconds to reach us on Earth and light this world. Finally, we get to use our natural stove! Nevertheless, why does it take such a long time for the light and heat to travel from the core to the surface of the sun? The reason lies in the path light takes to reach the Earth. Consider it takes a straight path from the outer surface of the sun to Earth, hence a travel time of just 8 minutes 20 seconds. However, inside the sun is a different story; the billions of internal reflections that light has to suffer in its journey from the core to the surface of the sun makes the journey too long, hence the 10,000 years to reach the surface!

Nature's own nuclear reactor, the sun deploys the fusion reaction to generate energy. We are not yet able to exploit fusion energy for any useful purposes. Scientists at the Lawrence Livermore National Laboratory has very recently announced a key achievement in fusion research. They focused laser light onto a target, which resulted in a hot-spot the diameter of a human hair, generating more than 10 quadrillion (1 quadrillion is equal to 1 followed by 15 zeros) watts of fusion power for 100 trillionths (1 trillion is equal to 1 followed by 12 zeros) of a second. Although we are still far away from exploiting fusion energy for commercial purposes, it is being considered a pathbreaking step towards the development of clean energy.

Though the sun is generous, it is the ultimate source of all our energy needs and indeed sustains all life on the Earth, it also sends blasts of charged electrons, protons, various ions as part of the solar wind. If unchecked, the solar wind can obliterate every form of life on earth. Luckily for us, the Earth's magnetic field protects us from the destructive solar wind.

THE MAGNIFICENT MAGNET: A SAVIOR

The solar wind is a stream of charged particles travelling at a very high speed, away from the corona or the upper atmosphere of the sun. This wind would have struck earth, had our Earth like a heavenly warrior not shielded its children, the living beings. This shield is a field created by an invisible magnet!

We have associated magnets with metals and alloys which can be magnetized. We know that magnets have an influence even outside their body. We also know that if we bring the north poles of two magnets close, they repel and the same happens for two south poles. Furthermore, when we bring a north pole of a magnet in the vicinity of a south pole of another magnet, they attract. Though there is no contact between the two magnets with just space between them, there is some sort of 'signal' between them which 'asks' like poles to move away and, unlike poles, to come close. This is the mechanism of action of our saviour, the invisible magnetic field of the Earth. The magnetic field repels and diverts the solar radiation from a distance from its original path, leaving the Earth unharmed.

Charles-Augustine de Coulomb, a French engineer, and physicist had discovered Coulomb's Law which describes the electrostatic force of repulsion and attraction between like and unlike electric charges. We know solar wind carries charge, electrons, protons, and ions; according to Coulomb's Law, other charges like or unlike should repel or attract them. It, therefore, appears strange that the Earth's shielding mechanism was guided by a magnet. Here we have to consider one of the most significant developments in our understanding of science. For long in history, people used magnetic bars and their interaction with the Earth's magnetic field to find direction. The earliest recorded reference and understanding of magnets were found in the works of the Greek philosopher and polymath Aristotle and Greek mathematician and astronomer Thales of Miletus. One of the earliest applications of the magnet, still in use found mention in the book Susruta Samhita by the Vedic Indian physician Susruta. He described procedures that involved the use of magnets to take out metal shards from wounds. Though magnets were used since ancient days, the properties of magnets were attributed to mysterious phenomena. It was not until 1600, when the book De Magnete published by the English physician and physicist William Gilbert, rejected the ancient theories of magnetism. The mysterious veil was lifted.

William Gilbert also studied static electricity. Static electricity also gave rise to similar forces between things carrying them but was distinguished from magnetism, as magnets had poles while those carrying static electricity did not have poles. The understanding of electricity was taken forward when the American polymath, Benjamin Franklin demonstrated the presence of charge in lightning. He even successfully showed that electric charge is quantized, meaning that charge can exist in discrete quantities, and there cannot be any charge in between these quantities. Coulomb gave an expression

> *In De Magnete, Gilbert studied static electricity produced by amber. Amber is called elektron in Greek and electrum in Latin. This gave rise to the modern terms electric and electricity.*

for the force between any two charges. Italian physician, physicist, and biologist Luigi Galvani observed that the legs of a dead frog twitched when an electric current was passed through them; this proved that electricity affects biological systems. Galvani's contemporary, the famous Italian physicist, and chemist, Alessandro Giuseppe Antonio Anastasio Volta, disagreed with Galvani about the source of the electricity. He argued that it was not animal electricity as Galvani had believed, but the source of the electric current was the metals and the conducting fluid in the tissues in the frog. Interestingly, this debate led to the development of the first battery, the "voltaic pile" by Volta.

VOLTA AND VOLTS

Alessandro Volta entered science with great potential. One of his earliest achievements was the discovery of the simplest hydrocarbon methane. He improvised a machine to demonstrate capacitance, the ability to store charge. A capacitor was the earliest form of electric storage, where a body was charged and retained the charge. Through further research and great innovation, the world's first electric cell, the primary component of the electric battery, was built. The voltaic cell also called the Galvanic cell, used two containers, one with a copper bar dipped in copper sulphate solution and the other with a zinc bar dipped in a zinc sulphate solution. The solution in both containers is ionized due to the polar nature of water. Copper being more electronegative compared to zinc as it has more affinity for electrons pulls an electron from the zinc bar when connected by a conducting wire. It was much later that values of electronegativity of different elements were known; one such was the Pauling scale developed by the great American chemist and biochemist Linus Pauling. Volta discovered that in his cell, the copper bar was pulling more charges and was getting thicker, as the positively charged copper ions from the ionized copper sulphate solution got attracted to the negatively charged copper bar, attached the electrons to themselves, and deposited on the copper bar as neutral copper atoms. Exactly the opposite happened with the zinc bar; the zinc atoms give up electrons, become ionized, and dissolves in the zinc sulphate solution as bivalent zinc ions. To make sure that the solutions do not get charged due to the release and absorption of ions, a salt bridge that maintained the charge balance by transporting the lost ions was introduced.

This discovery led to the further improvement of the battery. One such modern battery invented by Georges Leclanche was the Leclance cell. Since Volta, there has been enormous development in the design of chemical batteries. They can now deliver more power, and many are rechargeable. Modern batteries are powerful enough to store large amounts of charge to enable you to drive a car and be small enough to fit into a wristwatch hearing aid, or a cardiac pacemaker or last long enough to enable you to watch a movie on your smartphone. Most of today's modern electronic devices use the versatile Lithium-ion battery invented by the American materials scientist and solid-state physicist, John Bannister Goodenough and his team.

We cannot imagine a minute in our day without the battery. Volta's discovery has changed modern life beyond our imagination. His discovery will undoubtedly continue to shape the future of the world. It was not only the discovery of the battery but the very understanding of the nature of electricity that was transformed by Volta. Sir Humphry Davy, an English chemist, and inventor expanded the knowledge from the Voltaic or Galvanic cell and established the new subject of electrochemistry.

THE CHRONICLE OF CHEMICALS

Sir Humphrey Davy was initially a poet credited with writing about 160 poems, most of which were never published. He also wrote on metaphysics, geology, theology, and chemistry. He gained interest in science, especially in chemistry. He came across works of Lavoisier and Priestly. Influenced by

Davy experimented with nitrous oxide, astonished at how it made him laugh; he gave it the name "laughing gas" and wrote about its potential use as an anaesthetic agent during surgery. Davy also did pioneering work on the use of silver nitrate in reproducing photographs!

the works of Galvani and Volta, Davy pioneered the electrolysis of molten salts using the Voltaic pile. He discovered potassium in 1807 from the electrolysis of caustic potash and in the same year discovered sodium through electrolysis of caustic soda. Before him, no distinction had been made between sodium and potassium! Davy continued the electrolysis of alkaline earth (minerals) and discovered calcium, barium, strontium, and magnesium. Not done with his discoveries of new elements and continuing with his experiments, Davy isolated boron; a large number of elements, for an individual chemist indeed! It was a few decades later that Dmitri Ivanovich Mendeleev, a Russian chemist, and inventor, came up with the Periodic law, a scheme to scientifically list and classify the elements. He created the farsighted version of the periodic table of elements.

Armed with the knowledge of the chemical and physical properties of different elements, today, we are able to use them in a variety of applications, from nuclear reactors to aerospace. Predicting the property of elements was another miracle of the scientific method of

> *Mendeleev claimed to have got the structure of the Periodic table in his dream! Applying his Periodic Law, Mendeleev also predicted eight elements and some of their properties which were not discovered back then! He used the prefixes eka-, dvi-, and tri- from Sanskrit to honour the ancient Sanskrit grammarian Panini!*

listing them out. Before Mendeleev, Johann Wolfgang Dobereiner, John Newlands, and many others were on the lookout for a scientific method to list the elements then known. After Mendeleev, the English physicist Henry Mosely made it more scientific by introducing atomic numbers instead of atomic weights for sequencing the elements in the periodic table.

Swedish chemist Jöns Jacob Berzelius, who along with Robert Boyle, John Dalton, and Antoine Lavoisier, is regarded to be one of the founders of modern chemistry, was a contemporary of Davy. He was effusive in his praise of Davy's lecture on electrochemistry. Davy died at the relatively young age of 50 in Geneva, Switzerland but left behind his assistant and student, another person destined to change the world. Beginning as a bookbinder, he read books on science that came to him and was largely self-educated; Michael Faraday discovered the laws that governed electromagnetism and electrochemistry. Faraday established the concept of electromagnetic fields, the principles of electromagnetic induction and diamagnetism, and the laws of electrolysis! Faraday was one of the most prominent scientists in history. Physicist Ernest Rutherford speaking of Faraday said, *"When we consider the magnitude and extent of his discoveries and their influence on the progress of science and industry, there is no honour too great to pay to the memory of Faraday, one of the greatest scientific discoverers of all time"*. As a chemist, Faraday discovered the ring molecule of benzene, though its ring structure was discovered by the German organic chemist Freidrich August Kekule.

Kekule proposed the theory of chemical structure but was challenged by this molecule discovered by Faraday. Any mathematical understanding of its structure would not satisfy all of benzene's properties. Kekule had a daydream of a snake biting its own tail! From this, he got the idea of the ring structure of benzene, opening up a new world of organic ring compounds. This new understanding of ring compounds opened up a better understanding of it and aromatic compounds. Before we return to Faraday's other discoveries, let us see what benzene meant to chemistry.

THE CONSEQUENCES OF KEKULE'S DREAM

Kekule's cyclic structure of benzene was indeed a great innovation in structural chemistry. Kekule's structure was challenged by Albert Ladenburg, his former student but confirmed by Irish physicist Kathleen Lonsdale.

When we talk of benzene, we cannot possibly ignore its source, at least the most common source, crude oil. Crude oil extracted by drilling the earth is a source of thousands of organic substances; petrol, diesel, petroleum jelly, and many more were derived or synthesized from these components, such as plastics, synthetic fibres, so much so, we cannot imagine modern life without crude oil! One standard method of separating the various constituents of crude oil is fractional distillation. Every single thing we come across in our daily lives uses at least some component of these organic materials. That brings us to the question, what are organic materials?

Lavoisier divided chemistry into two branches; organic and inorganic. Organic chemistry dealt with substances or compounds derived from organisms, and inorganic chemistry dealt with all other substances and compounds. The Swedish chemist, Jöns Jacob Berzelius stated the law of vital force, which can be imagined as some divine force required to create an organic compound; it implied that organic molecules could not be synthesized in the laboratory! This was a widely accepted law till the German chemist Friedrich Wohler synthesized urea, a widely known organic compound, from starting materials that were inorganic in 1828! This discovery dealt a fatal blow to the doctrine of vitalism and revolutionized our understanding of chemistry. Armed with the knowledge of synthesis of organic compounds from organic and inorganic molecules created a new world where people could

> *Carbon is the 15th most abundant element in the earth's crust and the fourth most in the universe after hydrogen, helium and oxygen. It is also the second most abundant element in the human body after oxygen! It is the common element in all life form.*

synthesize their desired materials. Carbon differs from all other elements in its ability to form hundreds of thousands of compounds. What lies in the heart of all this is the property of catenation shown by carbon. Carbon atoms tend to form bonds with other carbon atoms almost endlessly. This amazing property of carbon is the reason that we have allotropes like graphite and diamond, which occur in nature and have a variety of uses, and even the strange soccer-ball-like structure in buckminsterfullerene, which is chemically C_{60}. Carbon's enormous ability to bond with hydrogen and many other elements is the motivation for a complete branch of chemistry dedicated to carbon compounds; organic chemistry.

Some of these organic compounds, such as hydrocarbons, are widely used as fuel. The most common source of carbon is coal. When Scottish inventor, mechanical engineer and chemist, James Watt invented his steam engine, which used coal as the source of power, it revolutionized transport and industry. Coal became the most important fuel to run power plants to cooking stoves. It was the beginning of the industrial revolution, which swept Great Britain and the rest of the world.

THE EXPLOSION THAT MOLDED THE WORLD

As the world started to grow small, with people colonizing nations; stepping where no one had ever stepped before; venturing into new lands, winning overall resistance, the human civilization proclaimed to everyone that they are the most powerful of the estimated 8.7 million species on earth.

Advances in medical sciences led to an increase in the survival rate for humans, which in turn led to unprecedented growth in population. As the population grew, so did the hunger for food, industries, transport, and everything else that made the human race more and more powerful. However, to satisfy that hunger, humans had to cross a big obstacle. The natural resources, coal and petroleum for energy, metals for machinery lay under big beds of rocks, and human muscles were too weak to break them and extract the resources. In the Bronze and Iron Ages, man extracted whatever was easily extractable. As the more easily available sources were exploited and depleted, it became harder and harder to obtain ores, and the demand also grew manifold due to advances in technology and rising incomes.

Nevertheless, the human mind would still not be at rest because, if they did not have powerful enough muscles, they had enough grey matter to break the rocks! As expected, the key to this problem was given by a human mind. The genius of Ascanio Sobrero, an Italian chemist, led to the invention of a black powder in 1847; it was nitroglycerine. When exposed to heat, even the slightest amount of the powder would cause an explosion strong enough to shatter a rock! The key to the apparently insurmountable problem was found. However, who would handle the angry powder! The desire was not fully achieved as its instability made it nearly impossible to use it on rocks as the slightest motion involved in carrying it caused it to explode. The key was found, but what to do with the key if the lock was missing!

Two decades passed by without any hint as to the opening of the lock. Twenty years since the invention of nitroglycerine, one of its manufacturers, was trying to figure out a solution for long. The story goes that the patent holder of 355 patents, the Swedish scientist, in whose name the most prestigious award in history is given, Alfred Bernhard Nobel, chanced upon the breakthrough when he accidentally dropped nitroglycerine on the ground. It so happened that contrary to expectations, it did not explode! Nobel immediately realized that the solution to the problem the human civilization was confronted with lay under his feet! To be more precise, in the soil on which he was standing. This soil contained diatomaceous earth. This earth, also known as Kieselguhr, absorbs nitroglycerine, hence making it more stable and controllable.

In 1867, twenty years after Sobrero's invention of nitroglycerine, Nobel invented the dynamite, a huge assistance to human civilization. Thereafter, Nobel continued his research and came up with many other explosives such as gelignite and ballistite. From then on, nothing stopped man from excavating mines, getting access to minerals, natural resources, fuel, and coal. They could use dynamite

and its successors to blast rocks and mountains and excavate tunnels through which underground rail or roadways were laid. The relentless industrialization of human civilization continued, this time with the help of dynamite.

INVASION OF THE MACHINES

With the advent of the steam engine and its increasing use in the wake of the Industrial Revolution, the challenge was to mine more and more coal. One of the challenges with coal and other minerals was the presence of water in the mines, which made the removal of coal and ores difficult. Initially, the water released in the mines was manually removed by buckets, *etc.* Even before large-scale coal mining was carried out using dynamite, Thomas Newcomen's pump working on an engine was used to pump out the water from the mines. This was the beginning of something called the steam engine. The steam engine works on the power supplied by steam generated by burning coal. However, Newcomen's engine used up large quantities of coal to mine coal, and a more efficient engine was needed. Many people worked on this, but a groundbreaking improvisation by the Scottish inventor James Watt changed everything. For a minimum of four centuries before the industrial revolution, the society and economy had largely remained static. Agriculture was the main economic activity, and the use of tools and machinery was rudimentary. People did know of some, such as the printing press invented by Johannes

Gutenberg during the renaissance. Nevertheless, the supply of essentials for survival, food, and clothing did not meet the demand.

Watt's improvisation of the steam engine ignited the fire of the industrial revolution ushering in new production processes in Europe and the United States. Watt's steam engine gave rise to the automated textile industry. His invention was exploited by George Stephenson, the "Father of Railways".

George Stephenson, unfortunately, courted controversy when he was accused of stealing the idea of the Safety Lamp from Davy! He was later exonerated, and the House of Commons committee in 1833 found that Stephenson had an equal claim as Davy for the invention of the lamp.

With his son, Robert Stephenson, who is often regarded as the greatest engineer of the 19th century, George developed the steam locomotive. Their rail transport system was one of the most significant technological inventions of the 19th century and one of the principal catalysts of the industrial revolution.

The development of the steam engine also revolutionized water transport. Ships, which depended on sails, now began to be fitted with powerful steam engines. The advancement that came with the industrial revolution was perhaps civilization's biggest leap. Cities began coming up where factories were built, and an increasing number of people started

The famous Titanic, the largest ship of the time carrying 2435 passengers and 892 crew could cruise at 39 km/hour with a maximum speed of 43 km/hour. It ran on 29 boilers feeding two steam engines with a output of 46000 HP.

working in factories. The cities lacked space for accommodating large labour forces which worked in the factories; slums started to grow. This was not limited to England and Scotland; the fruits of the industrial revolution spread to the whole of Europe, and with colonization and imperialism, Europeans spread to all the continents; North and South America, Africa, India, Australia. Fairy Queen, one of the oldest locomotives in the world built in 1855 at Leeds in England, reached Calcutta, India, in the same year. The railways in India rapidly expanded, as much for the governance of the British colony as for trade and commerce. Today Indian Railways is the largest employer in the country and eighth-largest in the world employing 1.254 million as of March 2020. Every country set up a specialized industry, maximizing benefits from their natural resources.

With industrialization emerged new problems. Businesses wanted to maximize profits; there was fierce competition between people, companies, provinces, and nations and in order to maximize profits, safety measures were compromised. Factory workers often suffered injuries and disabilities due to unsafe working conditions. Their health had enormous impacts on the quality and quantity of goods produced. All these and many more problems could not be addressed by common sense alone and needed innovative ideas.

The concept of an economy is as old and fundamental as the concept of society. However, in ancient times, the economics of a society used to function very differently. For instance, farming was the primary occupation; people used the barter system to trade amongst themselves, and so on. Modern-day economics as we study it came with Adam Smith, the "Father of Modern Economics". He introduced the concept of *"laissez-faire"*, or a free-market economy. Another path-breaking concept from him was the concept of division of labour which led to assembly line production. A subsequent major breakthrough in economics was brought about by Alfred Marshall, who introduced the fundamental concept of demand-supply equilibrium. He gave the laws of demand and supply that govern the market, introduced the demand-supply curve, and thus perpetuated the idea of market equilibrium, *i.e.,* stability between what the consumers demand and what the producers produce. The other important concepts put forward by him are those of marginal utility and marginal costs of production.

At the macroeconomic level, revolutionary economists like John Maynard Keynes made such remarkable contributions that a branch called "Keynesian economics" came up. His theory talked about the total spending in the economy on the whole and the effects variouschannels of spending can have on output, employment, and inflation; in the short run. Other notable

> *"In long run, we are all dead"- John Maynard Keynes. He advocated that "governments should spend money they do not have" to mitigate the adverse effects of recession and depression. His prescriptions were adopted by most western capitalist economies.*

contributions in macroeconomics were the Permanent Income Hypothesis by Milton Friedman and the Life-Cycle Hypothesis by Franco Modigliani. As the world became integrated, various trade models like the Ricardian model, Heckscher-Ohlin model, Specific-Factors model, etc., were also developed to guide international trade. Economics being an ever-evolving subject, newer and newer laws are being advanced every day. With these laws being developed over time, unimaginable control over the market has been gained, giving way to a better world.

THE MECHANICAL HORSE

The heart of the industrial revolution was the engine. But, what exactly is an engine? Newcomen's steam engine and Watt's steam engine both are heat engines that used steam to produce movement. The steam engines were later used to build locomotives that moved rail coaches and ships. These could carry cargo and a large number of people over long distances. However, the steam engine was not really suitable, for use in personal transport.

Bertha Benz, a German automotive pioneer and inventor, funded her husband, automobile inventor Karl Benz to build a carriage for her with which she could travel the countryside. In 1886, they presented the Patent-Motorwagen automobile to the world, working on a two-stroke engine. Their engine worked through piston movement with small explosions in a chamber caused by ligroin, a petroleum solvent in the presence of air. Bertha Benz is the first person in history to drive an automobile over a long distance. Her pioneering trip involved driving the Patent-Motorwagen Model III from Mannheim to Pforzheim, a distance of 106 km! From experience gained by her in this trip, she suggested some innovations such as leather brake linings and fuel line designs which were applied by Karl Benz. Many innovations were made to the car by many engineers. Henry Ford was able to make innovations for mass-scale production in an assembly line. But even today, the engine in the car works through explosions of a fuel-air mixture that push pistons and not by steam.

Engines were widely known due to the rapid expansion of factories, except it took the father of thermodynamics, the French mechanical engineer and military scientist, Nicolas Leonard Sadi Carnot, to provide a better understanding of the working of the engine. Although steam engines were known for almost a century before Carnot, there had been no scientific study of these engines. His theoretical

In 1824, at the age of 27 years, Carnot published his only book, Reflections on the Motive Power of Fire in 1824. He was a victim of a cholera epidemic, dying at the age of 36 years in 1832!

approach in analyzing engines revealed that it was not the fluid (steam) that mattered but the difference in the temperatures of the two reservoirs. This approach helped design more efficient engines industrially. Carnot's theoretical assumption of a perfect engine, called the Carnot engine enabled the great German physicist and mathematician Rudolf Clausius and British mathematical physicist and engineer William Thomson (Lord Kelvin) to establish the second fundamental law of thermodynamics and define the concept of entropy.

Engines are now better understood as any machine that converts any form of energy to mechanical energy. The concept of entropy that was developed has great significance in physics. It enabled the German chemist Walther Hermann Nernst to formulate the third law of thermodynamics which predicts that temperature can never reach absolute 0 or -273.15 degrees Celsius. He showed that as the temperature approached absolute zero, the entropy also approached zero. Another limit to physics and humankind that is imposed by nature.

KINETICS OF THE INVISIBLE

The laws of thermodynamics explained many things. But what explained thermodynamics? Let us take the simple case of piston movement when there is no explosion—the gas in a confined chamber with a piston fitted to it when heated pushes the piston up. French chemist and physicist Joseph Louis Gay-

Jacques Charles identified hydrogen as a suitable lifting agent for balloons. With the Robert brothers, Charles launched the world's first hydrogen balloon on 27th August 1783 from Champ de Mars, now the location of the Eiffel Tower. In the same year, they launched a manned hydrogen balloon.

Lussac discovered that when gases react at constant temperature and pressure, they do so in a constant ratio of volumes to the reactants and product. Gay-Lussac also formulated how gases tend to expand when heated, but he credited it to the unpublished works of French inventor, scientist, and balloonist, Jacques Alexandre César Charles. Charles' law establishes the relationship between the volume of a gas at constant pressure and its absolute temperature.

About 100 years before Charles, an Anglo-Irish natural philosopher, chemist, physicist, and inventor, Robert

Boyle, regarded today as the first modern chemist came up with his Boyle's law. This law, states that at a constant temperature when pressure is increased, gas contracts, and when pressure is decreased, the gas expands. In other words, there is an inverse relationship between the volume and pressure of a gas at a constant temperature.

These laws may seem obvious today, but back then, they were revolutionary; these laws enabled the connection of the highly debated microscopic particle theory of gases to their macroscopic

nature. A milestone connection was made between the two, microscopic theory and macroscopic nature, by an Italian nobleman who became a practitioner of law at the early age of 20 and soon found intellectual comfort in the study of physics and chemistry; Lorenzo Romano Amedeo Carlo Avogadro. Avogadro discovered that at the same temperature and pressure, an equal volume of all gases contains an equal number of molecules. This was a revolutionary concept as it applied to all gaseous elements or compounds, organic and inorganic, anywhere in the universe. This law gave birth to the amazing concept of mole, which enables us to calculate the number of molecules in macroscopic volumes of gases. A mole is the molecular mass of any element or compound expressed in gram; *e.g.* one mole of carbon would have a mass of 12 grams, and one mole of oxygen gas would have a mass of 32 grams, one mole of copper has a mass of 63.55 grams and so on.

The stage was set for the French engineer and physicist Benoît Paul Émile Clapeyron to combine all the above laws to formulate the ideal gas law; though it does not work for real gases. The real gas law was derived much later by Dutch theoretical physicist Johannes Diderik Van Der Waals, now known as Van der Waals' equation. However, ideal gas law was adequate for understanding the kinetic theory of gases.

Austrian physicist and philosopher Ludwig Boltzmann made a significant breakthrough and revolutionized physics. He used statistics and probability to provide a mathematical explanation of the ideal gas law and the second law of

Boltzmann modelled gas molecules as colliding billiard balls in a box. He argued that in the world of mechanically colliding particles, the second law was a result of the fact that disordered states are the most probable.

thermodynamics. Boltzmann is remembered for the development of the kinetic theory of gases and a statistical explanation for the second law of thermodynamics. Statistical mechanics, developed by him, is one of the pillars of modern physics. Unfortunately, Boltzmann committed suicide as a result of a mental depression now known as bipolar disorder while on vacation. In those days, there was no treatment for bipolar depression. Incidentally, he died in the same year that Spanish neuroscientist and pathologist Santiago Ramon Y Cajal, father of modern neuroscience, along with Italian biologist and pathologist Camillo Golgi, were awarded for their contribution to the understanding of the brain. This would eventually make the way for modern psychiatry.

Svante August Arrhenius, a Swedish scientist, received a grant to study with Boltzmann. Arrhenius made major contributions to ionic theories, the conductivity of electrolytes and offered definitions for acids and bases. Arrhenius was one of the pioneers of the science of physical chemistry along with Friedrich Wilhelm Ostwald, Jacobus Henricus van't Hoff and Walther Nernst.

The German physicist and physician Hermann Ludwig Ferdinand von Helmholtz made immense contributions to physical chemistry, chemical thermodynamics as also physiology and psychology.

The Maxwell-Boltzmann distribution, named after James Clerk Maxwell and Ludwig Boltzmann, is an outcome of the kinetic theory of gases; and provides probabilistic

> *The speed of sound in the air is about 343 m/s. The molecules in the air are therefore bombarding us at a speed which is one and a half times the speed of sound, incessantly!*

velocity distribution of gas particles (that is, atoms or molecules) in three dimensions at particular environmental conditions. From this distribution, the average speed of oxygen molecules around room temperature is calculated to be 500 m/s.

Josiah Willard Gibbs, an American scientist who made significant contributions to physics, chemistry, and mathematics, and about whom Albert Einstein remarked, "The greatest mind in American history" and Max Planck, "The greatest theoretical physicist ever", contributed significantly to the application of thermodynamics. Together with Boltzmann and Maxwell, Gibbs created statistical mechanics. The name statistical mechanics was coined by Gibbs. Gibbs also invented vector calculus, independently of the

English mathematician and physicist, Oliver Heaviside who too carried out similar work around the same time.

Apart from being a pioneer of statistical mechanics, Maxwell made huge contributions to the understanding of electromagnetic radiation. However, in order to understand it, we have to go back to Faraday and the Danish physicist and chemist, Hans Christian Øersted.

A SIGNIFICANT UNIFICATION OF HUMAN KNOWLEDGE

Hans Cristian Øersted believed that there was a relation between electricity and magnetism and had been looking for a connection between them since 1818. In 1820, he published his discovery that a current passing through a wire deviates a magnetic compass. Little did he realize what impact this observation would have on humanity! With more intensive investigations, he concluded that a current-carrying wire produces a circular magnetic field. Ørsted's work was a major step towards a unified concept of energy and stirred much research in the scientific community.

Born on 20th January 1775, André-Marie Ampère educated himself in the confines of the extensive library established by his prosperous businessman father, Jean-Jacques Ampère. He began teaching himself advanced mathematics at the early age of twelve. Though Jean-Jacques was called into public service of the revolutionary government, misfortune befell the family when he was guillotined at the time the Jacobin faction seized power during the French revolution. Ampère fell into depression for more than a year. He remained unproductive, but after his recovery, started teaching. Ampère wrote papers on topics ranging from mathematics, chemistry, philosophy to astronomy. When he got learnt of Hans Christian Øersted's discovery of the deflection of a magnetic needle by a current-carrying conductor, he began developing a mathematical and physical theory to understand the relationship between electricity and magnetism. This was the beginning of electrodynamics.

He devised an instrument with parallel wires which could move like a pendulum. He used the voltaic or galvanic cell to pass current through the wires. In his experiments with parallel current-carrying wires, he observed that when the

Ampère conceived the existence of the "electrodynamic molecule", the component elements of both electricity and magnetism. It was the precursor of the idea of the electron.

Ampère's name is one of the 72 names inscribed on the Eiffel tower.

current through the wires flow in the same direction, they attract and come close and when they carry current in the opposite directions, they repel and move further away from each other. His observations on the forces between the wires led to Ampère's law which is harmonious with Charles Augustin de Coulomb's law of magnetic action. Ampère thus explained the

Øersted's mystery of current diverting a magnetic needle. Ampère, applying his theory, developed a device to quantitatively measure a current through a conductor. A more advanced and modern version of such an instrument is called the Ammeter. Ampère thus laid the foundations of the new science of electrodynamics. The standard unit of electric current has been named Ampere in honour of his contributions to the understanding of electromagnetism.

Øersted and Ampère changed the way people looked at electricity and magnetism. Michael Faraday looked at the converse problem; can a changing magnetic field, in turn, cause an electric current to flow through a neighbouring wire. Faraday wrapped two insulated wires around an iron ring and observed that when a current is passed through one coil, a momentary current flowed in the other coil, the phenomenon of mutual induction. Ampère invented the solenoid, where a current is made to flow in a coil wound on an insulated cylinder, producing effects of a bar magnet. Faraday moved magnets back and forth through such a closed coil to show that current is produced in the coil. The change of magnetic field in the vicinity of a conductor generated current; this fact was used by Faraday to make something called the Faraday's disc, the earliest electric generator. Today every electrical device we use other than those which use single-use batteries, uses electricity from a power station. The most typical power stations are coal or thermal, nuclear, and hydropower station, which convert steam, nuclear or potential energy to electric energy with the help of the electric generator. Faraday also designed a couple of devices that produced what he called "electromagnetic rotation", in which rotatory motion was achieved due to interaction between a magnetic field and a current-carrying conductor, the first motor. A motor is a type of engine that converts electric energy to rotatory mechanical energy. The ceiling fan is perhaps one of the simplest applications of motors. Fans are installed in large factory sheds for circulating air, in laptops and mobile phones, where miniature fans cool the solid state processors, which get heated on use! There are thousands of other applications of generators and motors. In fact, without these electrical devices, human civilization can literally come to a standstill. The knowledge of unifying electricity and magnetism has enabled us to engineer and build machines that help us in nearly every field in every aspect of life. It gives us an insight into a reality that will forever change our understanding of everything. Yet now we know it was just the beginning of the unscrambling of the mystery of the universe, and what lies ahead cannot be imagined in a single day.

THE FOUNDATIONS OF MODERN TECHNOLOGY

On 30th April 1777, humanity was presented with one of the greatest minds that humankind has ever known. When he was very young, studying in primary school, everyone in the class was told to add all numbers from 1 to 100. This was to keep the children occupied. It was solved in seconds by a student to the astonishment of his teacher. It was Johann Carl Friedrich Gauss's grand entry into mathematics. What he did was absurdly simple; he added 100 to 1, 99 to 2 and so on; each sum was equal to 101. He then multiplied 50 by 101 and got 5050 as a result! This was the summation of an arithmetic progression. It was known long since the ancient Greek, Vedic and Egyptian era, but Gauss solved it independently at the age of 7! Gauss was a prodigy; at the age of 3, he corrected a mathematical calculation in his father's notebook.

Gauss would be remembered as one of the most influential mathematicians who ever lived for his groundbreaking contributions to many areas of mathematics and physics, including number theory, geometry, probability theory, geodesy, planetary astronomy and electromagnetism. Apart from many mathematical papers, Gauss produced quite a few students such as Peter Gustav Lejeune Dirichlet, Sophie Germain, August Ferdinand Mobius, Richard Dedekind, Gustav Robert Kirchoff, Bernhard Riemann, whose works were the foundations on which our civilization rests. We will come back to Gauss in later sections, for the time being confining ourselves to his contributions to electrodynamics.

To understand Gauss's law, one of the fundamental laws in electromagnetism, we need to understand flux. It is the number of field lines passing through a surface, a simple definition given by Faraday. We can visualize it better by considering, for instance, the electric field emanating from a charged body. Imagine an electric field that consists of vectors at every region in space. The direction of the vector will be determined by the type of charge causing the field; if it is a positive charge, the vectors will point away from the charge and if it is a negative charge, the vectors would point towards it.

The magnitude of the electric field can be calculated using Coulomb's Law for electrostatic force. Knowing the direction and magnitude of the field at every point, we can draw the vector at every point in space, and that would give us the field or vector field. We have earlier, come across vector fields in our understanding of fluid flow. Exploring the concept further, flux can be understood as the amount of such field coming out through any surface consisting of orthogonal components.

At an infinitesimal level, flux converging from all possible directions at a point or diverging from a point is what is known as divergence at that point. Gauss's law states that the flux of the electric field out of an arbitrary closed surface is proportional to the electrical charge enclosed by the surface. Stated differently, it states that the divergence of the electric field at any point inside a charged body is equal to the volume density of charge divided by a constant, called the permittivity of free space and outside the body is zero. This is Gauss's law of electricity and is equivalent to Coulomb's inverse square Law; one can be derived from the other. He applied the same relation for a magnetic field, where he arrived at something interesting. The value is always zero, which leads to the conclusion that for every magnetic flux outward, there is a flux inward; hence there cannot be a magnetic monopole (that is, a magnet with only one pole), unlike electric charges, which can exist as either positive or negative! Gauss's law of electricity was discovered before him by Joseph-Louis Lagrange. Keeping their works in mind and the contributions of Ampere and Faraday, we now visit Maxwell. From statistical mechanics to thermodynamics and now to electrodynamics, Maxwell made significant contributions, without which the development of this world we live in would not have been possible.

THE RACE TO BE THE SUPERIOR RACE

We believe that we are the rulers of the world. We are numerous, intelligent thinking animals, technologically advanced and have harnessed natural resources in a manner no other animals have done. Natural history has shown that humans are not the first to conquer the world and probably will not be the last. There have been other rulers before, like the cyanobacteria, which made other life possible because of their emission of oxygen which filled the earth with it. Much later came the massive living species of dinosaurs whose only living descendants, birds, remain. Other dinosaurs were victims of the Cretaceous-Paleogenic mass extinction event, which is attributed to a massive meteor strike by most researchers.

Humans are now not the only rulers of the world; the other massive territory is controlled by ants! Ants may form 15 to 25% of the terrestrial animal biomass. Nevertheless, how did the tiny ants conquer the world? Cyanobacteria had no competitor, dinosaurs were *The population of ants is about 10 billion billion; the human population was about 7.7 billion in 2019. More than 13800 of an estimated 22000 species have been classified.*

mighty, and so they could rule the world in their heyday. Like human societies, ant societies also display division of labour, communication between individual members and the ability to solve complex problems. Though they are not the only ones in the animal kingdom, their communication systems are more advanced than those of any other. Their elaborate communication system relies on chemicals they release, the pheromones and sound and touch. The pheromones help ants in keeping trail, as shown by Edward Osborne Wilson, the influential American biologist and naturalist widely regarded as "The Darwin of the 21st Century".

Humans, however, developed the most advanced communication that relies on languages and scripts. The invention of the postal system widened the range and speed of communication. We eventually invented electronic devices like the telegraph for a faster communication, which was developed and modernized by many scientists and engineers. American painter and inventor Samuel Finley Breese Morse invented the Morse code, a fast communication transmission language. Alexander Graham Bell, the Scottish-born scientist and engineer, invented the telephone. With these methods of communication, people were sharing their knowledge, information and were interlinked better than ever before. These enabled them to understand newer things through sharing of ideas which led to a faster pace of inventions.

Though this is not a world that we see. We live in the age of wireless communication and international connections across the seas. We cannot visualize a life without mobile phones and laptops. The entire humanity is connected through the internet. There have been unimaginable growth in communications technology and also phenomenal advancements in biology, such as knowing the shapes and working of proteins and DNA. We now know that the universe is expanding, we can "hear" the sounds from the Big Bang, and we also know why the fire glows. We can literally "look inside a rock" and know its chemical composition without

> *Ernest Orlando Lawrence invented the cyclotron, accelerating a particle to a high enough speed to bombard a nucleus, thus making it radioactive.*

damaging it. We have invented thousands of machines from the cyclotron to PET scan that enables us to observe the working of the brain and locating cancer cells.

JAMES CLERK MAXWELL

Despite major developments in physics, light somehow continued to remain a mystery. It was beyond the existing theories or experiments back then that could explain the nature of light. Though optics was a flourishing subject and people used the power of reflection and refraction to see the smallest to the farthest objects which were beyond the vision of the unaided human eye. However, there remained a question on the nature of light; was it a particle, or was it a wave? The first breakthrough was made by Isaac Newton when he split the rays of the sun into the visible spectrum using a prism. Newton theorized light as a flow of small particles, and this was known as the corpuscular theory of light. He presented this view in his treatise Optiks, published in 1704. Another leading scientist of the time, the Dutch mathematician and astronomer, Christiaan Huygens differed with Newton's theory claiming that light is a wave. Huygens assumed that light has a finite speed and offered the first mathematical theory of light. However, challenges remained, as, at that time, the wave theory did not explain the optical phenomena. This was the beginning of a great scientific debate that continued until 1801 till British polymath Thomas Young put an end to this problem. He passed light through a double-slit barrier and received an interference pattern on a screen behind it. This firmly established the wave behaviour of light. Slowly the nature of light began to be understood, but people still did not know how much potential it had to change science in theory and practice.

Born on 13th June 1831, the Scottish mathematical philosopher James Clerk Maxwell had an unquenchable curiosity from a very early age. From the age of 3, anything that moved, made a noise, or shone, fascinated him. Raised in his father's estate of about 1500 acres in relative

> *Maxwell was interested in psychology and investigated the perception of color, color-theory and color-blindless. This led him to apply his theory of color perception to color photography. In 1861, Maxwell presented the world's first demonstration of color photography using the principle of three-color analysis and synthesis.*

isolation, Maxwell did not fit well in school. Maxwell continued his studies, though he cared little about academics and the school syllabus. He won the first prize in school for English and poetry. At the age of 14, he wrote his first scientific paper in which he described a mechanical means of drawing curves with twine and also properties of ellipses and other curves. Maxwell also mathematically explained the stability of Saturn's rings, after working on the problem for two years. This was a problem that had eluded scientists for over 200 years! His conclusion that Saturn's rings were neither solid nor fluid but consisted of "brickbats" was confirmed through direct observation much later in the Voyager fly byes in the 1980s.

Maxwell was influenced by the brilliant work of English philosopher and chemist Henry Cavendish. He edited many of Cavendish's research documents and brought them to light. Maxwell married Katherine Mary Dewar, seven years senior to him. Katherine helped in his laboratory and was a partner in many of his experiments. Maxwell's contributions to physics are enormous, from the theory of colour vision to the first durable colour photograph to the development of dimensional analysis to kinetic theory of gases and thermodynamics!

Nevertheless, his most outstanding achievement was the formulation of the classical theory of electromagnetism, which brought together electricity, magnetism and light. These discoveries ushered in the age of modern physics and laid the foundations of fields such as quantum mechanics and special relativity. Many scientists regard Maxwell as the 19th century scientist who had the greatest influence on 20th century physics. Moreover, many consider his contributions to science to be of the same magnitude as those of Newton and Einstein. Maxwell died at the relatively young age of 48 years, the life of the genius cut short by abdominal cancer, a disease that did not have any cure back then.

Faraday and Ampere's work, the Gauss's two laws of electricity and magnetism, were all combined and mathematically formulated into twenty equations in twenty variables by Maxwell. Oliver Heaviside, the English mathematician and physicist, rewrote these equations into four differential equations, the modern form which we know *The growth of the radio industry following Hertz's discovery of radio waves changed the world. The radio, television, communication industry thus have their origins in Maxwell's publications.* as Maxwell's equations today. Maxwell revealed that the equations predict the existence of oscillating electric and magnetic fields that travel through empty space and that light is an electromagnetic disturbance that obeys the electromagnetic law. The French mathematician and physicist Jean d'Alembert gave the first wave equation, and Euler gave the first three-dimensional wave equation. Maxwell realized that the solution he obtained by solving his equations is the wave equation for electromagnetic fields. He even found that the velocity of propagation of the wave from his equation is approximately equal to the velocity of light. Light's mystery was unravelled and the light was henceforth known as an electromagnetic wave. It was also realized that light is not the only electromagnetic wave; it is just a small band visible to the human eyes in the whole electromagnetic spectrum. There are other waves we cannot see; these were experimentally proven through the discovery of radio waves by the German physicist Heinrich Rudolf Hertz in 1887, eight years after Maxwell's death. Maxwell's publications opened up a vast area of research and discovery. The already known infrared radiation was recognized to be electromagnetic waves. Wireless message transmission was demonstrated by Indian biologist and physicist Jagdish Chandra Bose with millimetre waves. The Italian inventor and electrical engineer Guglielmo Marconi demonstrated long-distance wireless transmission using radio waves.

Applications of electromagnetic theory and waves are inseparable from our everyday life. It has unlocked the gates to new areas of science and our understanding of nature, from the theory of space and relativity dealing with the largest to the theory of quantum mechanics, the theory involving the smallest!

THE SCATTERING OF ELECTROMAGNETIC WAVES

From Maxwell's equations, we come to know that light travels through a waving vector electric field moving perpendicular to a waving vector magnetic field. These fields are defined over time and space as any other wave. The electromagnetic waves move in a direction perpendicular to both magnetic and electric fields carrying the disturbance. But where are the electric or magnetic fields coming from? We have seen from Øersted and Faraday's laws that they each generate the other and go on forever until disturbed in the vicinity of another electromagnetic field.

Since the ancient days, people were puzzled at the change of colour of the sky from blue at noon to orange at dusk. The Italian polymath Leonardo Da Vinci had studied this problem and offered an explanation based on scattering. With an understanding of electromagnetic waves and their scattering, British scientist John William Strutt better known as Lord Rayleigh, came up with a theoretical explanation. The phenomenon is now known as Rayleigh scattering.

We know that the sky, for instance, is filled with gases

of the atmosphere and also minute particles that deviate the light waves. The earth is round in shape, and the angle of the interaction of the light waves with the particles is not the same everywhere. Moreover, white light consists of the whole visible spectrum seen together. Light rays with different wavelengths undergo deviations to different extents due to scattering. This results in a blue sky at noon, and due to earth's rotation, yellow or orange at dawn and dusk.

The German physicist, Gustav Adolf Feodor Wilhelm Ludwig Mie, described another type of scattering of electromagnetic waves by spheres, called the Mie scattering.

These scatterings are called elastic scattering because they work with the assumption that the kinetic energy of a particle is conserved in the centre of mass of the particle. In other words, there is no modification in the energy of the system. However, there exists a scattering where kinetic energy is not conserved in the incident particle in a system. Incident particle is what enters the system of scattering. Indian physicist Chandrasekhara Venkata Raman worked in the field of light scattering. He discovered that when light passes through a transparent media, some of the deflected light changes amplitude and wavelength; this new type of scattering of light came to be known as Raman scattering. American physicist Arthur Holly Compton in 1923, demonstrated the particle nature of electromagnetic radiation, which was a sensational discovery at that time. This was known as the Compton Effect.

Maxwell's equations, combined with the Navier-Stokes equation, gives birth to the subject, magnetohydrodynamics (MHD), less commonly known as electro-fluid-dynamics or electrokinetics; it deals with the study of electrically conducting fluids, such as plasma, liquid metals and electrolytes such as saline. Magnetohydrodynamics finds uses in engineering (plasma confinement, nuclear reactors, electromagnetic casting etc.), astrophysics, magnetic drug targeting *etc*. The sun is an MHD system that is not well understood.

The salty water flowing below the Waterloo bridge in London interacts with the earth's magnetic field to produce a potential difference between the two banks. Faraday tried, but could not measure this due to insufficiency of the instruments available in his time.

THE PERSPECTIVE OF PIERRE DE FERMAT

We have so far had an idea about what is light, some of its behaviour, explanation and uses. When we see things, it is because when light falls on them, some of the wavelengths in the visible spectrum get reflected. How do we see transparent objects then, like glass, gel, lens etc.? The great French

Pierre Fermat Last Theorem in number theory could not be solved for more than 300 years. Andrew Wiles finally solved the theorem in 1994. Fermat's correspondence with Blaise Pascal laid the foundations for probability theory.

mathematician Pierre de Fermat observed this and stated a principle, the principle of least time or Fermat's principle. To understand the principle, I would use a real-life situation. Imagine you have to reach a tree in a field for shelter from the rain, but a muddy patch surrounds the tree. If you want to get to the tree in the least possible time, you have to adjust the route in a way to optimize the time taken by walking a part on the better stretch and a part on the muddy one so that minimum time is taken. Fermat realized indeed it was the nature of light by demonstrating the way it bends when entering a glass or any other optically denser medium. Optically denser refers to the resistance to light diminishing its speed. This makes the light bend, and the resultant distortion causes us to see the glass or gel though they are transparent. It is because of this discovery that Fermat is considered an important figure in the historical development of the fundamental principle of least action in physics.

People soon realized that this principle does not explain the way any object with mass behaves. Though Newton's laws explained the motion of mass, a complicated system would require a more powerful version of Newton's laws to describe their motion. Say, ten magnetic bars are placed on a table, and iron dust is sprinkled. What would be the trajectory of the iron particles to reach their respective destinations? Even if we assume that the experiment is carried out in a vacuum, the attraction and repulsion of all the ten magnets make it a tremendously difficult problem to figure out which of the infinite possible paths the iron particles would actually take. French mathematician Pierre Louis Maupertuis and German diplomat and polymath Gottfried Wilhelm Leibniz independently developed the principle of least action, also known as Maupertuis's Principle. Leonhard Euler gave a mathematical formulation to this principle. This principle was essentially Newton's Second Law of Motion, but it gave a new way to analyze the mechanics of every object in the universe. The principle is central in modern physics and mathematics and has been applied in fluid mechanics, thermodynamics, theory of relativity, quantum mechanics, particle physics and string theory! The principle states that the path an object will take with be the path having the least action. Action is the value of the potential energy of the particle subtracted from the kinetic energy of the particle at that point. The path that the particle takes is the path where the summation of action at each point is the least among all paths.

This principle was powerful. Italian mathematician and astronomer Joseph-Louis Lagrange used this in the formulation of Lagrangian mechanics. With Lagrangian mechanics, it would be much simpler to calculate the path of iron particles in a field produced by several magnets.

However, modern science that has slightly different principles, works better with the Hamiltonian mechanics derived by great Irish mathematician and astronomer William Rowan Hamilton. In quantum mechanics as well as statistical mechanics, the Hamiltonian approach is generally preferred.

In all of these approaches, to sum up, actions at all the points in the path, we need to add up infinite values. How do we do that?

THE LANGUAGE OF THE GODS

In the age of dinosaurs, a volcanic eruption took place and formed hard igneous rocks. Continental drift took this part of the land and joined with what we now call India. The people inhabiting India thousands of years ago carved huge caves in these rocks and even sculpted them. Nevertheless, the sculptors and architects of antiquity, how did they know which part of the rocks to sculpt and carve and not risk the demolition of the Ellora caves? Much before, the Egyptian pyramids were built with heavy blocks of rocks in a manner that the weight distribution of the rocks did not lead to the collapse of the structure into the interior cavities of the pyramids.

The great Greek mathematician Archimedes arrived at the volume and surface areas for cones and spheres. Those were the early days of mathematics. None of the modern tools was available back then, yet these seemingly implausible tasks were carried out. What was the mystery? Little fragments of techniques worldwide were coming together to give birth to a branch of mathematics now called calculus. It is an arcane branch of mathematics originally about geometrical shapes, which has reshaped human civilization; it is everywhere around us; without calculus, we would not have cars, ships or aeroplanes; computers, mobiles, or television! The American physicist Richard Feynman famously remarked about calculus "It is the language God talks." Not one person can be credited for the development of calculus; it is a collective achievement of thousands of mathematicians and scientists over many centuries!

Modern calculus was developed by Isaac Newton and Gottfried Wilhelm Leibniz. There was rivalry between them; they independently laid the foundation of calculus as we know it today. Of course, there were many ideas since the ancient days in Greece and later in China, India and the Middle East, which guided the development of concepts

Newton gave his calculus the name "the science of fluxions", Leibniz gave the name calculus. Most of the modern notations (e.g. d/dx) were also introduced by Leibniz.

that have later come to be included in the broader branches of integral and differential calculus. Leibniz and Newton had different approaches, but each was able to formulate the concept; Leibniz began first with integration while Newton with differentiation. It was indeed an international effort; French mathematicians

Augustin-Louis Cauchy, Rene Descartes and Pierre de Fermat, English mathematician Brook Taylor, German mathematicians, Johann Carl Friedrich Gauss and Georg Friedrich Bernhard Riemann, Swiss mathematician Leonhard Euler and many others have created calculus. The amazing powers of calculus are still being expanding and refined.

To understand calculus in a very basic sense, let us consider a Cartesian plane such as a simple graph sheet. The Cartesian coordinate system and the Cartesian plane were described by Rene Descartes and Pierre de Fermat independently. Any point in this plane can be described by a set of two values which give the perpendicular distance of that point from two fixed mutually perpendicular axes, the x-axis and the y-axis. The study of geometry using this Cartesian plane is Cartesian coordinate geometry. Any curve in this coordinate system can be defined on the basis of the location of any point on the curve, each such point being defined by a set of two numbers, one the x-coordinate or the abscissa (the perpendicular distance of that point from the y-axis) and the y-coordinate or the ordinate (the perpendicular distance of that point from the x-axis) This value of ordinate can be defined as some function of x or the abscissa. The function here is just a mathematical equation; it may refer to a curve with a curvature, or a straight line or some discrete points on the Cartesian plane. Physically, the slope of a curve refers to the rate of change of one value (say the ordinate) with respect to the other (the abscissa). Differential calculus deals with infinitesimal parts of the curve, so small that the part can be considered as a point and the slope of the curve at that point. The more the absolute value of the slope, the steeper the curve is at that point. To calculate the slope of the curve is differentiation. The slope of the curve can also help us find if there is a maximum or minimum value of the curve and where. It also enables us to quantify the curvature or steepness of a curve. Differentiation has its limitations, though. Imagine you run your finger along a curve from the left side and another from the right side. If your fingers do not meet at some point, then the curve is discontinuous at the point. Differentiation cannot be carried out in such a case at there is no unique slope at that point. Even in a continuous curve, differentiation may not be possible. Say we consider a triangle without its base; we cannot differentiate at its tip or its vertex, though the "curve" is continuous at the vertex. At the vertex, the steepness cannot be determined as it is inclined both to the left, and to the right at that point. Differentiation gives the slope of a curve, but why is slope important? Because, slopes of many curves, give physical quantities of interest to us. For example, if a curve gives the position of a moving car with time, then the slope would give its speed, and if a curve gives the speed of the car with time, the slope would give its acceleration at any point in time.

The German mathematician Karl Theodor Wilhelm Weierstrass defined a curve, which has no discontinuity but still cannot be differentiated at every point, as the curve has vertices. Many real-life curves are Weierstrass functions. Examples of such curves include fractal curves, which we encounter in Brownian motion. When a function depends on several variables; an example being a surface on a three-axis Cartesian coordinate system, differentiation can be applied against any variable, that is against any axis. This takes us to the realm of partial differentiation, which determines the changes in the value of a function against a particular variable with other variables not undergoing any change.

ADDING UP ALL ACTIONS

When we can differentiate a function, can we do the converse? Can we find out a function, by differentiating which, we can get the function we started with? It can be achieved by a process called integration. But why would we do that? Newton-Leibniz theorem, also known as the fundamental theorem of calculus, tells us that if we integrate a function over a range, we obtain the area under the curve made by the initial function. The German mathematician, Georg Friedrich Bernhard Riemann further explained that if we approximate the area under a curve by making small rectangular partitions for simplicity, we can estimate the area of each rectangle. If we add the area of the rectangular partitions, we will get an approximate area under the curve. However, if we decrease the width of the rectangular partitions and increase their number, our approximation will be more accurate. By increasing the number of partitions

Riemann made a hypothesis, which has a connection to the famous Goldbach conjecture, which states that any even number greater than 2 is a sum of two prime numbers. This remains unproven to date. Riemann's ideas provided the mathematical foundation of Einstein's theory of general relativity.

(conversely by decreasing the width of the rectangular partitions), the area under the curve calculated as the sum of the areas of the partitions will be closer and closer to the real area. If we make an infinite number of partitions, our estimate of the area is with infinitesimally small error and is the real area under the curve in a given range. Let us go back to the problem of calculating the path of least action for a moving point mass. Lagrange assumed point masses and defined action to be dependent on position and velocity at the point (equivalently, potential energy and kinetic energy at each point). Velocity is just the incremental change in position against incremental change in time. Therefore velocity can be attained by differentiating position with respect to time. The position also happens to be dependent on time. Action, therefore, is a function of time and therefore carrying out integration for a given range of time, we can determine the path with the least action. From Riemann integral, we know that integrating was just adding up action along the whole curve at all infinite possible points.

With calculus, solutions appear like magic. Moreover, the magic of calculus is visible in nearly every area in physics, mathematics, computing, medicine, engineering, mathematical economics, statistics, chemistry, biology and countless other areas of human knowledge. More practical results are obtained when differential values are substituted in place of variables in equations involving polynomials, such as in Newton's second law or the wave equation.

Polynomial equations, though, is another thing that requires a deep understanding of mathematics. It is not only deep but also strange.

EXPONENTS AND COEFFICIENTS

Multiplying 2 by itself gives us 4. The square root of 4 gives us 2. Multiplying - 2 by itself gives us 4. What will the square root of 4 give us now?

There are variables with different powers in polynomial equations, and they have arbitrary constant values attached to them (that is, coefficients). Let us consider a polynomial equation with a single variable with different powers. The polynomial would be known by its degree that is the highest power of the variable in the polynomial equation. If we draw the curve represented by this equation on a Cartesian plane, at some points, it may intersect the x-axis; these are called the zeroes of the function. Many mathematicians have worked with such polynomials, and all the approaches show us that the degree of the polynomial equation is equal to the number of its zeroes. This would imply that if we have a polynomial equation with degree 10, there would be 10 zeros of the equation. When the variable is substituted with any of these ten values, also called roots, the value of the equation would equal zero. From here we can conclude that the equation $y = x^2 - 4$ has two zeros;

that is, for x = + 2 and x = − 2, the right-hand side of the equation is equal to zero. This is what is called the fundamental theorem of algebra. It indicates that the cube root of 1 must have three values. In other words, three numbers exist whose cube is 1 (the relevant equation is $y = x^3 − 1$). One of these roots is natural (1), and the other two are complex numbers. Complex numbers are numbers that have two parts; a real part and an imaginary part. The fundamental imaginary number is root over − 1 (that is $\sqrt{−1}$), represented as i. It can be thought of as the number squaring which gives us − 1.

However, there is a problem. How do we get the roots of a polynomial equation? For equations with degree 2, we have the famous quadratic formula. The solution of quadratic equations was known since ancient times. The Greek mathematician Euclid in his book Elements describes a geometric method to solve such equations. Another Greek mathematician, Diophantus, in his Arithmetica, provides an algebraic method. The Indian mathematician Brahmagupta provides a formula equivalent to the modern form in his book, 'Brahmasphutasiddhanta'. For equations with degree 3, we have the gigantic cubic formula. There is a formula for the roots of a degree 4 polynomial equation as well.

A question mathematicians were bothered with was how to derive a formula for roots of polynomial equations with

higher degrees, such as 5, 6, *etc.*, and how far could one go? The Italian mathematician and philosopher Paolo Ruffini claimed that we could not solve quintic and higher-order polynomial equations using the approach for quadratic and cubic equations. Ultimately, it was Norwegian mathematician Niels Henrik Abel who gave the first complete proof that there cannot exist any general algebraic solutions to polynomial equations with degree 5 (quintic) or above. While doing this, Abel invented a branch of mathematics called the group theory, independent of the French mathematician, Évariste Galois. He sent it to Carl Friedrich Gauss, who, unfortunately, discarded it as "the work of a crank." It was understood better when Évariste Galois, laid the foundations for Galois theory and group theory, two major branches in abstract algebra. Unfortunately, Abel died at the age of 26 from tuberculosis, and Évariste

Galois died at 20 fighting a duel! Galois's collected works run into only 60 pages, yet they contain important ideas which have far-reaching consequences for almost all branches of mathematics! Abel and Galois did not live long, yet the fundamental contributions they made to many branches of mathematics is immortal.

CALCULATING BEYOND LIMITATIONS

There were these limitations of mathematics, and then there were limitations of the human mind. Equations were created and solved, but in practical applications, it was difficult to calculate large numbers and amounts. Since the ancient days, man has searched for a mechanical aid to calculations. One of the world's first calculators, the abacus, was used in Europe, China and Russia, centuries before the Arabic numeral system was adopted. Blaise Pascal eventually invented the mechanical calculator in 1642, followed by Leibniz, who invented a digital mechanical calculator, which used the Leibniz wheel. The French inventor Charles Xavier Thomas de Colmar, patented the Arithmometer, the first digital mechanical calculator, it became the

> *Wilhelm Schickard, a German professor of Hebrew and astronomy, had described a calculating clock in his letters to Johannes Kepler, twenty years before Pascal. Nevertheless, the machine was not complete.*

> *Babbage directed the construction of some steam-powered machines which could calculate. For ten years, he received government funding to the extent of £ 17000, but then the Treasury lost confidence in him!*

first commercially successful mechanical calculator. Simple arithmetic calculations could be carried out by the arithmometer, but analytical and more complex computing required something more powerful. The English polymath and inventor, Charles Babbage often regarded as the father of the computer, originated the concept of the first digital programmable computer. He developed the first mechanical computer, the Difference Engine, that could operate on different algorithms. He proposed a successor to this design, the Analytical Engine. The Difference Engine could compute values of polynomial functions. Though Babbage's machines were mechanical and unwieldy, their basic architecture was similar to modern computers; the data and program memory were separate, the operation was instruction-based, the control unit could make conditional jumps, and they had separate input/output units.

The daughter of poet Lord Byron and Lady Byron, Augusta Ada Lovelace was an English mathematician primarily known for her work on Babbage's proposed mechanical

> *Ada Lovelace dismissed Artificial Intelligence. She wrote, "The Analytical Engine has no pretensions whatever to originate anything. It can do whatever we know how to order it to perform. It can follow analysis, but it has no power of anticipating any analytical relations or truths."*

general-purpose computer, the Analytical Engine. She created an algorithm for calculating a sequence of rational numbers called Bernoulli numbers. This algorithm is regarded as the first computer program and Lovelace the world's first computer programmer!

The English mathematician, computer scientist, cryptanalyst, philosopher and theoretical biologist Alan Mathison Turing is widely regarded as the father of theoretical computer science and artificial intelligence. He formalized the concepts

John Von Neumann could divide two eight-digit numbers in his head when he was six years old. He could converse in Ancient Greek. He knew Hungarian, French, English and Latin. He was a master in Byzantine history. By the age of eight, Neumann was familiar with differential and integral calculus!

of algorithm and computation with the model of the Turing machine, which can be deemed to be a model of a general-purpose computer. The Hungarian-American polymath, physicist, computer scientist, mathematician John Von Neumann proposed the architecture of the modern computer system, where both the data and program are stored in the same address space; it is commonly called the von Neumann architecture. Evidently, computer science, a formal field of science, was taking birth.

Sir John Ambrose Fleming, an English electrical engineer and physicist, invented the first thermionic valve or vacuum tube, a significant step forward in the "wireless revolution" and an invention that is often regarded as the beginning of modern electronics. Fleming's diode was employed in radio receivers and radars for many decades. The vacuum tube was used to control electric current and was used in the first electronic general-purpose digital computer ENIAC (Electronic Numerical Integrator and Computer). ENIAC was a programmable computer. The ENIAC was designed by John Muchly and J Presper Eckert of the University of Pennsylvania and was primarily used for calculating artillery firing tables. Its first program was the study of the feasibility of the thermonuclear weapon! But ENIAC was unwieldy, weighing 27 tons and required 1800 square feet of space! By the end of its operation, ENIAC had 18,000 vacuum tubes, 7,200 crystal diodes, 1,500 relays, 70,000 resistors, 10,000 capacitors, and approximately 5,000,000 hand-soldered joints! And this giant consumed 150 KW of electricity!

The computational world got a major impetus when Austro-Hungarian physicist and electrical engineer Julius Edgar Lilienfeld identified field electron emission and invented the field-effect transistor (FET). The transistor in its modern practical form was invented by American physicists John Bardeen,

> *A semiconductor material is one which has an electrical conductivity falling somewhere between an electrical conductor such as copper or aluminium and insulators, such as glass and wood. Unlike metals, their resistivity falls with increase of temperature. Popular semiconductors are silicon, germanium and gallium arsenide. The conductivity of silicon is increased by adding minute quantities of elements like the pentavalent antimony, phosphorus and arsenic or trivalent, boron, gallium and indium; a process called doping.*

William Shockley and Walter Brattain in 1947. The first transistor used a crystal of germanium with gold contacts. The transistor could work as an amplifier; a small signal applied to one pair of its terminals can be used to control a much larger signal between another pair. However, the avatar of a transistor as an electrically controlled switch, in which current in a circuit can be turned off or on, by applying a signal to one terminal, is of immense interest in the computer world. Thousands of such transistors were wired to make a computer. Computers from the ancient hand-held abacus to Pascal's calculators to Leibniz's Arithmometer to Babbage's Difference Engine had come a long way. The journey to the modern laptops, mobile phones and supercomputers of today has been a long one and was made possible through the contributions of many scientists and engineers. As transistor technology

improved, they became smaller and smaller and could be tightly integrated into small semiconductor chips. This saw the exponential rise in the power of computers and a simultaneous decrease in their size. Today a mobile phone we use has more computational power than the combined computational power of all the Allied Forces in the Second World War! This explosive technological growth was summarized by the American businessman, engineer and founder Chairman of the Intel Corporation, Gordon Earle Moore, in his Moore's Law. The law forecast that the number of components (including transistors) in integrated circuits would double every two years. Over the years, the development of integrated circuits followed the Moore's Law. This prediction has become a target for miniaturization in the semiconductor industry. The earliest products using transistors were transistor radios which typically had 4 to 8 transistors.

The development of computer science and technology at a rapid pace gave rise to the Digital Revolution or the Third Industrial Revolution in the latter half of the twentieth century. The Digital Revolution marked the commencement

> *The Industrial Revolution saw the introduction of machines for mass production of goods in Europe and the United States between the years 1760 to 1820 and 1840.*
>
> *The Second Industrial Revolution also known as the Technological Revolution referred to that phase which saw rapid standardization and industrialization and is generally considered to have lasted between 1870 to 1914.*

of the Information Age, suggesting sweeping changes brought about by digital computing and communications technology, for creation, storage, transmission and retrieval of information.

Three other inventions; the monolithic integrated circuit chip by Robert Noyce at Fairchild Semiconductors, the first successful metal-oxide-semiconductor field-effect-transistor (MOSFET) by Mohamed Atalla and Dawon Kahng at Bell Labs, and the complementary MOS (CMOS) process by Frank Wanlass and Chih-Tang Sah at Fairchild paved the way for higher levels of integration of transistors. These were followed by the development of the silicon-gate MOS by

Federico Faggin, a Fairchild engineer; which he used to develop the first single-chip microprocessor, the Intel 4004. This laid the foundation of the microcomputer revolution of the 1970s when the home computer was introduced. One of the earliest microprocessors, the popular Intel 8085, had about 6500 transistors integrated on a single chip.

Parallelly, technology for interconnecting computers took shape. In the 1960s, the first message was sent over the ARPANET (Advanced Research Projects Agency Network), the first wide-area network of computers set up by the Advanced Research Projects Agency of the United States Department of Defense. Many such packet switched networks were developed around that time and in particular, the ARPANET led to the development of rules for connecting different computer networks; the precursor to the World Wide Web.

The English computer scientist, Sir Timothy John Berners-Lee invented the World Wide Web in 1989 which was available only to universities and governments at the time. It became publicly accessible in 1991. But to access information on the web, one needed a web browser. American software engineer and entrepreneur Marc Lowell Andreessen and Eric

Bina, introduced Mosaic, the first internet browser which helped in the development of the later browsers, Netscape Navigator and Internet Explorer.

In 1994, the Stanford Federal Credit Union was the first financial institution to offer online internet banking services to its members. Internet became a part of mass culture by the middle of 1990s and by 1999, almost every country had a connection.

The development of the MOS and large-scale integration (LSI) technologies and advances in information theory and cellular networking led to the development of affordable mobile communication. In 1973, American electronics engineers, John Francis Mitchell and Martin Cooper, demonstrated the first handheld mobile phone; it weighed about 2 kilograms! In 1979, Nippon Telegraph and Telephone (NTT) launched the world's first cellular network in Japan. The mobile phones became as ubiquitous as computers by early 2000. Though text messaging existed in the 1990s; they became popular only in the early 2000s.

The explosive growth of computer usage and mobile phones led to storage and transmission of information through digital media, on a scale never seen before. It is believed that in the late 1980s while less than 1% of the world's technologically stored information was in digital format, it had grown to 94% in 2007 and was more than 99% by 2014! With the increase in demand for digital storage, the capacity to store information has seen a phenomenal growth from 2.6 exabytes (1 exabyte = 1 billion gigabytes or 1000 petabytes) in 1986 to around 5000 exabytes (or 5 zettabytes) in 2014.

The phenomenal growth of the digital environment can be appreciated from the following figures:

Year	Cell Phone Subscribers		Internet Users	
	Number	Percentage of then world population	Number	Percentage of then world population
1990	12.5 million	0.25%	2.8 million	0.05%
2000	1.5 billion	19%[*]	631 million	11%[*]
2010	4 billion	68%	1.8 billion	26.6%
2020	4.78 billion	62%	1.54 billion	59%

[*] The percentages are calculated on the estimated population of 2002.

Computers are ubiquitous today. Every sphere of human activity is touched by it; modern life is impossible without the aid of computers, mobile technology and the internet. Whether it is booking a movie ticket, or a flight or a train ticket or ordering groceries or accessing one's bank account, it is the computer with internet that are at play. At the workplace, it is essential for financial accounting, human resource management, inventory handling and marketing. In engineering, computers are used for process control, and design of complex machine parts (Computer Aided Design) and machining (Computer Aided Manufacturing) of these parts and simulation of complex processes. One cannot think of complicated operations such as placing a satellite in its orbit, crunching terabytes of data to forecast the weather or cyclones, or simulating nuclear explosions without the aid of very powerful supercomputers. Computers, mobile phones along with internet are the backbone of social media, such as WhatsApp, Facebook, Instagram and Twitter which helps people to stay connected across continents.

All computers draw on the work of the English mathematician and philosopher George Boole. The self-taught Boole developed Boolean logic, which laid the foundations of the information age of which we are a part. Boolean algebra defines the logic operations involving "yes" and "no" (equivalently '0' and '1' or 'true' and 'false'), the building blocks of all computer operations and the language that allows us to communicate with the inanimate computer. Transistors are integrated into integrated circuits (ICs), the building blocks of all computers; they fortuitously operate in a manner that can easily mimic the logic operations in Boolean algebra. However, there are extensive codes within us and machinery, a mysterious and far more powerful computer created by God.

CODES OF THE CREATOR

This computer, which is very much within us, codes for nearly every structure and function in us. Evolution is a long term process, but in the short term, a species is well maintained due to these codes. We can easily recognize a human child, as they inherit nearly every structural and functional characteristic of their parents, who happen to be humans as well. However, there can be some general differences, first studied in the pea plant. It was observed that some pea pods are wrinkled and some are not; some plants grow tall, while others remain short, yet all of them resemble their parent plants. These were questioned by a farmer's son, who in his childhood practised gardening and studied beekeeping. The Austrian mathematician and biologist Gregor Johann Mendel went on to study mathematics and physics. His teacher in physics was the Austrian mathematician and physicist Christian Doppler. However, Mendel could not succeed in his examinations for becoming a high school teacher. He became a monk, in part because he could not afford to pay for his education. In his monastery's experimental garden of two hectares, Mendel cultivated some 28,000 plants, most of them, peas. Over seven years he studied seven traits that were passed on to the succeeding generations. From his observations, Mendel formulated three laws, namely the Law of Dominance, Law of Segregation and Law of Independent Assortment, which came to be collectively known as Mendel's Laws of Inheritance! Unfortunately, scientists of his time could not grasp the importance of his findings. The profound significance of Mendel's work was recognized much after his death; he was recognized as the "Father of Genetics" posthumously. Even Charles Darwin, who worked on evolution, was not aware of Mendel's work!

While the German biologist, Walther Flemming, unaware of Mendel's genetics, was investigating rapidly dividing cells, he found something is present in the cell nucleus, which gets distributed into the nuclei of the two daughter cells. These structures

> The Science Channel named Flemming's discovery of mitosis and chromosomes as one of the 100 most important discoveries of all times and one of the ten most important discoveries in cell biology.

strongly absorbed dyes; he named them chromatin. This process repeats itself. Flemming named it mitosis from the Greek word *"mitos"* for thread. Based on his discoveries, Flemming, for the first time, surmised that all nuclei came from a predecessor nucleus. He identified that chromatin was related to the threadlike structures in the cell nucleus – the chromosomes.

American geneticist and physician Walter Sutton and German biologist Theodor Boveri independently noticed that, as chromosomes multiply and divide at the cellular level, they too obey the Mendelian laws of inheritance. This is often referred to as the Boveri-Sutton chromosome theory. Finally, genetics and chromosomes were connected through the pioneering work on fruit flies (*Drosophila melanogaster*) by American evolutionary biologist and geneticist Thomas Hunt Morgan. Morgan was a prolific writer with 22 books and 370 papers to his credit. While crossbreeding fruit flies, he noticed that most fruit flies have red eyes, but some had white. He also noticed that ones with white eyes were mostly males, while ones with red eyes could be either male or female. Unlike Mendel's pea plant, fruit flies were not hermaphrodite, and therefore Morgan concluded that some of the traits showed sex-linked inheritance. This conclusion also proved Sutton-Boveri theory that traits are passed from generation to generation through chromosomes.

We have seen that if the same chromosome pairs that carry the genetic material from the parents are passed onto their children, the brothers and sisters should have similar features. Contrary to theory, no one was identical. Morgan theoretically explained that there must be some mixing taking place within chromosomes.

It took another great mind, American cytogeneticist, Barbara McClintock, to discover the phenomenon of genetic recombination by crossing over – a mechanism by which chromosomes exchange information. This crossing over is what leads to children of the same parents to possess variations though sminor. This crossing over led to the evolution of diverse species. McClintock also showed how different types of cells are formed from the same chromosome materials, from blood cells to neurons.

Outstanding contributions from Morgan and McClintock and many others established the subject of genetics. They threw light on how chromosomes behaved like computers that ran programs for particular species to come into existence. Genetics is very intensively linked to evolutionary science. But what is the molecular basis of this phenomenon?

DECODING THE TWINED THREAD

A nearly two-meter-long thread condensed into a nucleus with a diameter of about a millionth of a meter is what the chromosome is like. This thread, the deoxyribonucleic acid or DNA coils and supercoils to form 23 pairs of chromosomes, in the case of humans, to divide without any break in the thread during cell division. This gigantic molecule, DNA, dictates to the cell what to do and how to do it. From the codes of DNA, several proteins such as the enzyme helicase are produced, which works on the DNA itself.

Swiss physician and biologist Johannes Friedrich Miescher isolated the hereditary material from cells, the nucleic acids, DNA and ribonucleic acid, in short, RNA. Ribonucleic acid is another essential material for heredity, but these were well understood only after their structure was comprehended.

Using X-ray diffraction imaging, English chemist and X-ray crystallographer Rosalind Franklin discovered the key properties of DNA. A contemporary British biophysicist, Maurice Wilkins, obtained clear X-ray images of DNA. These discoveries enabled British molecular biologist, biophysicist and neuroscientist Francis Harry Compton Crick and American molecular biologist, geneticist and zoologist, James Dewey Watson to work out the double-helical structure of DNA. DNA is a long polymer molecule looking like a twisted ladder; with every step representing a bond between two nucleotides. The nucleotides are monomeric units consisting of nitrogen-containing nucleobases, a sugar, called deoxyribose and a phosphate group. There are four types of these bases, namely, cytosine, guanine, adenine and thymine. The steps of the ladder consist of a cytosine base connecting with guanine and the adenine with thymine according to a base-pairing rule. Both the strands of the DNA store the same biological information through the sequence of the four nucleobases. This structure enables the replication of the DNA when the strands separate as the sequence in one strand reveals the complementary sequence in the other strand, which is essential for DNA multiplication without distorting the code.

For copying a paper with text or image, we use a photocopier. Nature also has to use a photocopier to allow the offspring to be copies of their parents. Nature's photocopier is incredibly sophisticated. DNA plays a central role in carrying the codes for the synthesis of a particular protein. These proteins play a very decisive role in living organisms, as enzymes,

Frederick Sanger devised a method of determining the DNA sequence rapidly, which is known as "Sanger sequencing". They used this method (dideoxy method) to sequence human mitochondria DNA (16,569 base pairs) and bacteriophage λ (48,502 base pairs). This technique was eventually used to sequence the entire human genome.

hormones or structural proteins. Many outstanding scientists have worked to solve the mysteries of the genetic code. Some of them are the South African biologist Sydney Brenner, American biochemist and geneticist, Marshall Warren Nirenberg, Indian American biochemist, Har Gobind Khorana, British biochemist, Frederick Sanger, German biochemist, J Heinrich Matthaei, American biochemist, Robert William Holley.

DNA is like a factory, playing a central role in the synthesis of proteins. The double helix DNA unwinds and opens up into two single strands. Every individual base in the opened-up DNA attaches to a complementary RNA base. This is the single-stranded messenger RNA or mRNA, which carries a sequence of the bases complementary to the sequence in the DNA and enters the cytoplasm. The mRNA interacts with the transfer RNA or tRNA that attaches itself to the mRNA strand. There are three bases in the tRNA that fits against a particular region in mRNA. One end of the tRNA gets attached to the mRNA; at the other end of the tRNA, amino acids join consecutively to make a bond. There are 64 combinations of 4 bases (4 x 4 x 4) in 3 slots of mRNA and therefore in the tRNA too. Each such combination in the mRNA is called a codon, and its complementary anti-codon on the tRNA, which brings the particular amino acid and the beginning and end command of the protein chain. This protein chain, with amino acid sub-units, wraps up in a particular shape due to molecular forces and forms the final protein molecule or protein unit.

Ultimately, all the characteristics of an organism, are coded in these DNA in the nuclei. And these are passed on from the parents to their offspring. In all organisms, information from one generation to the next is passed on through the chromosomes. In humans, all the information that is passed on is contained in the 23 pairs of chromosomes. Sections of each DNA strand, known as genes codes for all traits of each human being; the colour of his eyes, the colour of his skin and hairs, his height, his intelligence, his structure and so on. Most traits in humans as also all organisms, result from a combination of environmental and genetic influences. Advances in molecular genetics helped us to understand human genetics and the mechanism of transmission of traits in humans. For example, some genes on the Y chromosome are passed from father to sons while some others are inherited from the mother. A complete understanding of the genes in a human organism would require the complete sequencing of the genetic content of the chromosomes, also known as the human genome. It was a gargantuan task, involving sequencing of over 3 billion base pairs in the human chromosomes!

DECODING THE HUMAN JIGSAW PUZZLE

The base sequences of many human genes had been determined through the individual efforts of many scientists. But a complete sequencing of the human genome would require huge funding and collaboration between many scientists and institutions. Many questions were raised on the merits of carrying out this sequencing, the risks it involved and the huge costs. Finally, in 1990, the Human Genome Project was initiated under the leadership of the American geneticist Francis Collins with funding support from the US Department of Energy and National Institutes of Health. Soon scientists and institutions around the world joined the effort. The project got traction due to advances in the techniques for sequencing of genes and developments in computer hardware and software which allowed the rapid analysis of the huge data generated. In 1998, Celera Genomics, a private sector company headed by American biochemist, J Craig Venter, also got into the race independently of the government-sponsored project. The competition turned to cooperation when the two projects joined hands which speeded up the decoding exercise. In June

2000, Collins and Venter announced the rough draft sequence of the human genome. After further analysis and refinement, the final sequence was announced in April 2003, completing the HGP although some gaps still remained. The completion of the Human Genome Project coincided with the 50th anniversary of the publication of the double-helical structure of the DNA by Watson and Crick.

Scientists have estimated that there are 22300 protein-coding genes and about 3.1 billion base-pairs in the human genome. The HGP has led to explosive advances in genetics and genomics. One such effort is the International HapMap Project which aims to identify similarities and differences in human populations in the continents of Asia, Africa and Europe. Comparative genomics is the study of similarities and dissimilarities between different species. Huge databases of full genome sequences of different organisms have been created and are now being used in comparative genomics. The sequence of DNA is stored in huge databases available to anyone on the internet. These studies throw fresh light on the evolution of species and genomes.

The availability of the complete gene sequence has opened new vistas in medical science. Genes have been identified which are associated with some diseases, for example, certain types of cancer; this allows accurate diagnosis of diseases even before the onset of clinical symptoms! The HGP has also revolutionized forensic science through DNA sequencing from samples obtained at crime scenes. Interestingly, the advances in genome sequencing technology have also followed Moore's law (which talks of levels of integration in integrated electronic circuits). These advances have found applications in many fields such as agriculture, animal husbandry, biofuels, bioprocessing, bioarchaeology and evolution. The commercial development of DNA based products is now a multi-billion dollar industry.

TOOLS THAT HELP US TO LOOK INTO GENES

The German mechanical engineer and physicist Wilhelm Conrad Röntgen had discovered X-rays in 1895. Unscrambling the structure and functions of protein, DNA, RNA required knowledge of how X-rays were diffracted by crystals. X-rays are scattered by crystalized molecules, and the interpretation of the scattering provides the leads to the crystal structures. The British father-son duo of William Henry Bragg and William Lawrence Bragg developed the analysis of crystal structures through the use of X-rays; the technique came to be known as X-ray crystallography. Lawrence, inspired by Agatha Christie's work, aspired to be a detective. His deduction of crystal structures from the patterns of the X-ray diffracted by crystals was no less startling! The American physicist Arthur Compton had discovered that X-rays scattered by electrons had longer wavelengths than the incident X-rays. Waves are usually represented by graphs depicting disturbance in the form of displacement from a mean position with respect to time. However, waves are often interpreted by transforming the time-distance graph of the wave disturbance into frequency-amplitude representation through a transformation called the Fourier Transform. The French mathematician Jean-Baptiste Joseph Fourier had a century earlier investigated the Fourier series, which eventually developed into the Fourier analysis and Fourier transform. The Fourier transform is a powerful mathematical transform widely used in many branches of physics, such as spectroscopy and signal processing.

CHAOS AT THE MOLECULAR LEVEL

From X-ray's interaction with the structure of DNA, RNA, proteins, and other molecules to determination of what force shall be necessary to break existing chemical bonds and to generate new chemical bonds are phenomena that happen practically at the molecular level. Each cell in our body "knows" which part of its DNA is to react in order to form the specific protein template RNA. We have seen how complex each operation involving protein synthesis is; however, this is just one phenomenon. In reality, there are billions of billions of activities happening in and out of us to maintain what we are! It is almost chaos there, yet there is some order.

Austrian physicist Erwin Schrödinger proposed that genetic information was contained in the configuration of the covalent chemical bonds. This, at the time when the DNA structure had not been decoded, was a very promising idea. Though he was a physicist, his works turned out to be of foundational importance to chemistry and biology. German physicist Walther Ludwig Julius Kossel and American physical chemist Gilbert Newton

Lewis' observation of noble gases made them conclude that electrons in atoms in molecules want to attain the configuration of noble gases. To get a better picture, let us go back to light, Maxwell's electromagnetic wave theory and black body radiation.

The German physicist Gustav Robert Kirchhoff and some others were studying radiation and spectroscopy. He coined the term blackbody radiation for an ideal physical body that absorbs all incident electromagnetic radiation irrespective of wavelength or angle of incidence. It is called "black-body" as it absorbs all colours of light. Lord Rayleigh and English physicist and astronomer James Hopwood Jeans solved the problem of blackbody radiation and established a law, now called Rayleigh-Jeans Law. This law gives the energy of electromagnetic radiation at different wavelengths radiated by a black body (which does not reflect electromagnetic waves) at a given temperature. This law predicted that at shorter wavelengths, the emitted energy would tend towards infinite. While the predictions by the law agreed with experimental results for large wavelengths, they broke down at shorter wavelengths (called ultraviolet catastrophe). The law was based on classical physics of the time, and observations revealed a crucial error in the interpretation then. Classical physics explained many things and continues to do so, but for blackbody radiation, clearly, some other vision was required. The answer to this problem led to the development of a new understanding of physics.

A brilliant student and gifted musician chose to study physics. However, his professor recommended he not study physics, as he believed that in physics, "almost everything was already discovered!". He replied that he only wished to understand the known fundamentals in the field. Thus began the journey of Max Karl Ernst Ludwig Planck in the realm of physics. The German Bureau of Standards had commissioned Planck to make light bulbs more efficient. Electric bulbs was a growing industry then, and it made immense sense to make them generate maximum light for the least electrical power. While working on this problem, Max Planck came across the Rayleigh-Jeans law. He was convinced that there was some wrong assumption in that law. He came up with a revolutionary idea that all electromagnetic energy radiated by black bodies are in discrete quanta, each containing an energy $h\upsilon$ proportional to the frequency υ (Greek small letter "neu"), the constant, h, called the Planck's constant. He found that with this new idea, experimental results fit in with the theoretical conclusions. The implication of Planck's radiation law is that at higher temperatures, total radiated energy increases and the intensity peak of the radiation shifts to the visible spectrum. Seen from a classical viewpoint, there is an infinite number of wavelengths possible, so energy would add up to infinite; which is not what is observed. If the energy is certain multiples of Planck's constant, the problem is solved. German theoretical physicist Albert Einstein used these multiples, what Planck called quanta and showed that once they strike a metal surface, electrons would be emitted; thus, the quanta behaves like a particle. The Danish physicist Niels Henrik David Bohr adapted Rutherford's nuclear structure to Planck's quantum theory and created the Bohr model of the atom. From Maxwell's equations, it can be derived that any accelerating particle with charge will experience a change in energy, and therefore electrons ought to collapse into the nucleus; Bohr's model explains why electrons orbit in certain energy shells or orbits of the given radius and also explained why the moving electron that happens to be electrically charged does not emit energy due to acceleration and does not collapse into the nucleus.

The publication of Max Planck's works heralded the birth of quantum mechanics, which revolutionized the understanding of atomic and subatomic processes. Max Planck is regarded as the "father of quantum mechanics". Quantizing light is a fundamental concept in modern or non-classical physics.

QUANTUM LEAPS

Like any other well-formulated theory, quantum mechanics was a result of the work of many geniuses put together. Many fundamental achievements radically changed our interpretation of the world around us.

Max Planck's proposition, that electromagnetic radiation from blackbody has discrete energy levels, indicating its particle nature, and Einstein's suggestion that light could behave as a particle encouraged the brilliant French physicist Louis Victor Pierre Raymond de Broglie to give a radical hypothesis. He not only proposed but mathematically justified why there is no priority of only light to

> *"Everything we call real is made of things that cannot be regarded as real. If quantum mechanics hasn't profoundly shocked you, you haven't understood it yet." – Niels Bohr*

behave both as particle and wave. He began with his revolutionary theory of electron waves that is electrons behaved both as particles and as waves and extended the argument to any matter or particle. With this idea, he was able to explain Bohr's theory, why electrons were confined to orbits of certain energy levels. Louis de Broglie proposed that the particle with momentum p should exhibit wave properties with wavelength λ, which would be equal to Planck's constant divided by its momentum (that is, $\lambda = h/p$). This was a very successful prediction and perhaps one of the weirdest. Later experiments, however, proved the validity of the prediction. This led to a strange situation, a particle, such as a sub-atomic one, is at once a particle at a point as well as a wave!

Newton's laws were accurate when observed at the classical scale, say for our everyday objects or celestial bodies. However, when dealing with the behaviour of matter and waves at the atomic and sub-atomic scales, we require an entirely different physics, known as quantum physics. Austrian theoretical physicist Erwin Rudolf Josef Alexander Schrödinger, considering de Broglie's wave-particle duality described the behaviour of such systems by a wave equation, known as the Schrödinger's equation. This equation is the pillar of quantum mechanics and has information relating to a particle, embedded in it. The solutions to this equation are wave functions which provide the probabilities of occurrence of physical events. This equation helps us in understanding the complicated mechanics that take place in chemical reactions, even those within our bodies.

The interpretation of Schrödinger's equation needed another great mind, the German physicist and mathematician, Max Born. The fundamental question was if everything is waving, what is it waving on? Light particle or photon was waving on electric and magnetic fields, but what about fundamental particles like an electron? The solutions to Schrödinger's equation, the wave functions were complex-valued numbers that cannot be interpreted in our natural world. Max Born formulated the probability density function from the real and imaginary parts of the wave functions. It tells us the probability of finding a particle at a given point and time in space. However, it was a deep understanding, as it gives us the probability. If we pry into this probability by observing the particle, the whole wave function collapses!

The solutions of Schrödinger's wave equations are wave functions that could only be related to the probable occurrence of physical events. In Newtonian mechanics, the planetary orbits are definite and visible, while

Schrödinger became interested in Indian philosophy, specifically in Upanishads and Advaita Vedanta's interpretation. He said that "There is obviously only one alternative, namely the unification of minds or consciousness. Their multiplicity is only apparent, in truth there is only one mind. This is the doctrine of the Upanishads."

in quantum mechanics, solutions are replaced by the more abstract notion of probability. Many physicists, including Schrödinger, were very unhappy with this aspect of quantum physics. Indeed, Schrödinger himself spent much of his later life formulating objections to the theory he himself had helped create! In a thought experiment devised by him, a cat is locked in a steel box with a radioactive substance. The mechanism is such, that after an hour, there is an equal probability of one atom of the substance either decaying or not decaying. If the atom decays, a contraption releases a toxic gas that kills the cat. However, until we open the box and the atom's wave function collapses, the atom's wave function is in a superposition of two states; decay and non-decay. Thus the cat in the box is in a superposition of two states; alive and dead! Schrödinger considered this outcome to be quite ridiculous. The fate of the cat has been a matter of much debate among physicists. This is the famous 1935 thought experiment Schrödinger's cat.

THE DISPARITY IN DISTANCES

These interpretations were getting bizarre. If you see a tree in front of you, you would expect it to be there irrespective of whether you are looking at it or not. But an implication of quantum mechanics is that if you close your eyes, meaning you are not observing the tree, the tree could be anywhere! The German theoretical physicist Werner Karl Heisenberg derived one of the most famous interpretations, the Heisenberg uncertainty principle or indeterminacy principle. The

Heisenberg admired oriental philosophy and saw parallels between it and quantum mechanics. After a conversation with the Indian poet Rabindranath Tagore on Indian philosophy, Heisenberg stated, "some of the ideas that seemed so crazy suddenly made much more sense."

uncertainty principle proposes that the position and velocity of any object cannot be measured precisely at the same time. The consequence of this is that the concept of exact velocity and exact position does not hold in nature! Mathematically, the principle states that the uncertainty of position multiplied by the uncertainty of velocity of a body is equal to or greater than a minuscule constant, $h/4\pi$, where h is Planck's constant. This seems to be counterintuitive as one could say that it should be easy to locate a moving car and measure its velocity accurately at the same time. The uncertainties involved with large objects are too small to be observed, however. At the atomic and sub-atomic levels, however, the product of uncertainties become significant. This implies that if we try to determine the position of a sub-atomic particle accurately, say an electron: it would be disturbed in such a way that measurement of its velocity would be inaccurate. This is not an insufficiency of the measuring instrument, the observer or the technique; it is a fundamental truth of nature. It is an outcome of the wave-particle duality of matter. So, the next time you see a person in front of you, consider him to be like a mist! You cannot exactly locate the person and cannot simultaneously accurately locate where they will be in the next moment!

The English theoretical physicist Paul Adrien Maurice Dirac derived a new equation, the Dirac equation, which predicted that for every matter, there is antimatter! Like for a negatively charged electron, there is a positively charged antielectron or positron, with all properties identical except the charge, which is positive. Dirac predicted the presence of this particle in 1931; one year later, positron was inadvertently detected in cosmic rays by the Americal physicist, Carl Davis Anderson. When matter and antimatter meet, they exterminate themselves, radiating gamma rays. The Positron Emission Tomography or PET scan, a valuable tool applied for the detection of cancer, makes use of gamma rays emitted during the interaction of positrons emitted by certain positron-emitting atoms and electrons. The strange part is, where is this antimatter, and why is it not annihilating everything we know to exist around us? Where is this missing world?

The interpretations of quantum mechanics are strange, but luckily it does not affect our observations in this macroscopic world. Their phenomenon is prominently observed in the world measured near the Planck scale. Max Planck, after quantizing the energy of a photon, went further to define natural units. How small can something be? Let us say time; we can visualize a second or even half a second. We can keep splitting time into half, and then again half of that and so on; intuitively, there should be no problem. Quantum mechanics showed that there is a limit to which time can be split.

The Planck time is the shortest time, which cannot be further divided. A value of time shorter than that has no meaning in this universe.

Similarly, Planck length is the smallest value of length with meaning in this universe. We are safe from quantum strangeness as they occur near these extremely small quantities. If we take the Hubble scale of the universe, which is to measure the largest at one end and Planck length as the smallest, at the other end, a living cell would be in the region of the average of the two scales. For reference, 1.616255×10^{-35} m is the Planck length. Light travels 299792458 meters in a second. Hubble length is 14.4 billion times the distance light travels in a year. In this universe, we are simultaneously extremely small and extremely large. The predictions of quantum field theory and general relativity are not expected to apply at the Planck scale.

An example of this is that our understanding of the Big Bang, some 13.8 billion years ago, does not extend to the Planck epoch, that is when the universe was less than one Planck time or 10^{-43} seconds old. In the Planck epoch, the currently established laws of physics may not apply. Some of the fundamental Planck units are given below:

Name	Value in SI unit
Planck length	1.616255×10^{-35} m
Planck mass	2.176434×10^{-8} kg
Planck time	5.391247×10^{-44} s
Planck temperature	1.416784×10^{32} K

The gigantic universe gives us reasonable evidence, once again, to comprehend how mysterious nature is.

RELATIVITY AT THE MACRO SCALE

Galileo Galilei observed that when we observe or describe any motion, it is always relative to another object. He argued that in a ship, moving straight on the sea without acceleration, we have no way to say it is moving unless we observe the surroundings. This thought experiment assumes no sea waves are present. Galileo also gave a mathematical transformation that relates the space and time coordinates of two systems moving at an unchanging velocity with respect to one another. He thus created the basic principles of relativity that the laws of physics are the same in any system that is moving in a straight line at a constant speed. This principle provided the basic framework for Newton's Laws of Motion. It was also central to Einstein's Special Theory of Relativity.

> *Albert Einstein called Galileo the "father of modern science." Galileo has also been called the "father of scientific method" and "father of modern physics."*

On another side, ground-breaking theories were developing on the motion of celestial bodies.

Mathematician and astronomer C l a u d i u s P t o l e m y ' s g e o c e n t r i c model of the universe

(with the earth as the centre of the universe) was being questioned by the Renaissance polymath Nicholas Copernicus and others who formulated the heliocentric model of the universe (with the sun as the centre of the universe). The Danish nobleman Tycho Brahe's observations of the planetary motions and his innovations enabled him to get better data than ever before. Brahe's assistant and German mathematician and astronomer Johannes Kepler worked on these observations and discovered the three iconic laws of planetary motion. Kepler's laws greatly influenced Isaac Newton, who discovered the law of gravitation, a specific interpretation of Kepler's third law. Newton used his laws of gravitation and motion to derive Kepler's laws and showed that planetary motion could be explained using physics and mathematics. One of the greatest polymaths of all time, Pierre Simon Marquis de Laplace discovered innumerable phenomena of celestial mechanics; his invariability of planetary mean motions led to the establishment of the stability of the solar system.

The complete understanding of relativity and celestial mechanics were pivots to an understanding of the complex working mechanism of the universe. However, relativity took a significant turn after the development of electromagnetic theory. When Maxwell's equations of moving charges and induced magnetic field are put in the equations of Galilean transform, the predictions do not match the observations. There was something missing, either in Maxwell's equations or in Galilean transformations. German-born American physicist Albert Abraham Michelson and American chemist Edward Williams Morley collaborated in the iconic Michelson-Morley experiment to detect the velocity of earth with respect to the luminiferous ether, the medium which was believed to carry light waves. They discovered something paradoxical in terms of classical Newtonian physics; the velocity of light was invariant irrespective of the velocity of the observer! Galilean transformations could not explain this. The great Dutch physicist Hendrik Antoon Lorentz proposed the magical, unintuitive set of equations, the Lorentz transformations which predicted that the length of a moving body and time with respect to it is not invariant and depends on the velocity of the body. Lorentz concluded that when a body travels at velocities close to that of light, its mass increases, it contracts in the direction of motion, and the time dilates! All of these are completely counterintuitive.

It took the great mind of Albert Einstein to fully comprehend and independently propose the constancy of the speed of light. Einstein's reasoning led to the new equations for space and time; and provided a deeper understanding of the Lorentz transformations predicting length contraction and time dilation. Einstein also concluded that simultaneity of events is relative; that is, events that are simultaneous for one observer may not be so for another observer; a concept that goes entirely against the predictions of classical mechanics. Einstein's theory of special relativity changed our understanding of the universe. The observation of subatomic particles moving at speeds close to that of light has confirmed the predictions of special relativity.

It is said that Mileva Marić, Einstein's first wife and his fellow student had collaborated in the development of special relativity. Einstein also rediscovered through a different approach the most famous equation in physics,

$E = mc^2$. This equivalence was initially proposed by the French mathematician and theoretical physicist, Henri Poincaré, whose approach was difficult to comprehend. This equation provided the great unification of mass and energy in physics.

MASS BENDS LIGHT

The implications of Lorentz's and Einstein's work was that if a stationary observer and a moving observer observe the same speed of light and the speed of light is a constant, then the value of time must be different for the two observers. This led to the concept of time dilation. Since the light has to travel the same amount for the stationary and moving observer, the moving observer needs to contract to make up for the relative reduction of light's movement. These are some of the amazing interpretations of the theory of special relativity.

German mathematician Hermann Minkowski developed a model for space with four dimensions, three of them relating to physical space (say x, y and z) and the fourth being time, called Minkowski space. This was a pillar on which Einstein built the Special Relativity and General Relativity.

Einstein, in a thought experiment, assumed a box accelerating upwards with a beam of light entering through a hole on one side and exiting through another hole on the opposite side. An observer inside the lift would see that the path taken by light inside it is a curve, in fact, a parabola. Pierre de Fermat had shown that light takes the shortest path. This suggested that in an accelerating environment, a straight line may not be the shortest path between two points. Einstein proposed the ground-breaking idea that gravity bends light and the curved path is the shortest path as the space gets curved. He said any mass that accelerates something towards it behaves like the accelerating box. He concluded that the presence of massive bodies curve space-time. The English astronomer, physicist and mathematician, Arthur Stanley Eddington's astronomical observation of positions of stars during a total solar eclipse proved this theory.

In this supermassive space, there can be a supermassive mass that can completely bend light towards itself. These bodies of extremely intense gravity do not allow anything to escape them, not even light, the enigmatic black hole. The structure of a black hole can be calculated using the General Theory of Relativity; the singularity consists of the centre of the black hole, a point of zero volume and infinite density and a surface that hides it, the event horizon. Inside the event horizon, the escape velocity of matter exceeds the speed of light. Since nothing can move at a speed larger than the speed of light, nothing can escape the event horizon! However, the English theoretical physicist Stephen William Hawking showed through the application of general relativity and quantum mechanics that black holes do emit subatomic particles, and they too decay when all their energy is exhausted!

The Einstein field equations are a set of ten nonlinear partial differential equations in four variables that relate space-time to the distribution of mass and energy within it. One solution to these equations leads to the conclusion that the universe is expanding. Einstein introduced a constant called the cosmological constant to make the universe stable; in other words, it is neither expanding nor contracting! However, modern astronomical observations reveal an accelerating expansion of our universe. The solutions to the Einstein Field Equations sometimes give rise to solutions that are contrary to common sense. For example, some solutions give rise to universes in which one can time travel into the past! This solution leads to the logical contradiction that allows a person to go back into the past to murder his parents before his birth!

THE BIG BANG CONUNDRUM

The Russian physicist and mathematician Aleksandr Aleksandrovich Friedmann argued that the universe is expanding and is not static, implying that all distant galaxies were moving farther and farther away from each other and from the centre of the universe. Belgian priest, astronomer and cosmologist, Georges Lemaître proposed the idea that if the universe is expanding, then it must have been expanding from the past. If we go on calculating backwards, we find no existence of space. He theorized that out of this void, a sudden burst of space began to form, the Big Bang theory took birth. A question that arises is: what was there before the Big Bang? We have come back to Minkowski space, where time is itself a part of the four-dimensional space-time. Before Big Bang, there was perhaps no time. Space, time, mass, energy, forces all emerged from mere nothing!

When Lemaître proposed the Big Bang and an expanding universe, astronomers and physicists, including Einstein, were sceptical. It was still a theory until American astronomer, Edwin Powell Hubble proved that the universe was indeed expanding with his observations of faraway galaxies. He noticed that far away stars are redder, meaning they emit light of larger wavelengths. The Austrian mathematician and physicist Christian Doppler had shown that the observed wavelength of a wave depends on the relative speed of the object and the observer. Applying this to electromagnetic waves emitted by stars, the red light emitted by far away stars indicates that the stars and distant galaxies are moving away from earth. Hubble proved one of the most radical theories of science, that the universe is expanding. Gravity pulls, yet everything is moving away; how is that so? If space itself is expanding, then what is it expanding into?

One of the earliest electromagnetic radiation from the Big Bang, the cosmic microwave background radiation, fills the whole space. These radiations are the residual effect of the Big Bang 13.8 billion years ago, a visitor from the event of the birth of the universe!

We have so far traversed the spectrum of what is perceptible by our senses, from the minutest to the enormous expanse of the cosmos. Of course, at the extremes of the scale, we are not able to directly 'see' the objects such as the subatomic particles and viruses or visualize the vastness of the universe. Even if not directly perceptible by the human senses, the ingenuity of the human mind, clever instrumentation and above all, revolutionary theories help us visualize the universe around us. The universe is complex and incredibly huge; the world of sub-atomic particles is another curious world where particles behave like waves and vice versa, and there are uncertainties about their position and velocity! Yet, no matter how complex these worlds are, the human mind has unravelled the mysteries piece by piece as if in a giant zig-saw puzzle since the earliest days of human civilization. And this has been possible due to what scientists believe to be the most complex thing known in the universe; the human brain. Whatever we know

> *Our galaxy, the Milky Way, is possibly between 100,000 to 150,000 light-years across. But the universe is far larger; it is estimated to be 93 billion light-years in diameter. One light-year is about 9.46 trillion kilometres, or more exactly, 9,460,730,472,580,800 metres!*

> *On the other end of the scale, a bacterial cell averages) 0.5   5 x 10^{-6} metres in length, while a virus typically has a diameter between 40   400 x 10^{-9} metres! The coronavirus is about 120 x 10^{-9} metres in diameter. An atom is 10^{-10} metres in diameter, while the diameter of a nucleus is just 10^{-15} metres!*

about the universe, whatever we perceive and interpret, are in our brain, an infinitely complex biological computer far more sophisticated and powerful than any computer built by man.

EVOLUTIONARY ENIGMA

We feel touch, and we smell perfumes, we hear music, and we see the infinite variety of objects and colours around us. All sensations, all feelings, all meanings are in our brain. The brain is the centre of all sensations and all instructions that allow all our movements. Apart from the sensory and motor functions, our brain has a vast repertoire of different capabilities. We display a wide range of emotions, have motivations, and can innovate and take logical decisions. Humans are creative in art, literature and visualization. This brain is what gives us, humans, evolutionary superiority over all other creatures. Human curiosity and its ability to perceive, analyze, and innovate have led to the remarkable development of human civilizations from the days of the caveman.

The curious nature and ability to possess such intense love and attachment are all a result of chemical signalling in our brain and the nervous system. When a signal is transmitted through a nerve, it travels by means of an ion-exchange impulse flow. The basic cell of the nervous system, the nerve cell or neuron, consists of the cell body or cyton, nerve fibre or axon, and the receiving processes or dendrites. Specific ions enter the axon and exit throughout the length of the axon. The movement

of these ions across the wall of the axon and into a neighbouring axon is what constitutes the transmission of signals. Once the signal is transmitted or received, the original condition of the particular region of the axon is restored through chemical processes. Whether or not to promote this impulse flow across the individual brain cell depends on the concentration of ions coming in and out. At the end of the axon, specific chemicals are released, which fits the receptor protein of the adjoining neuron, just like in our lock and key model for delivery of hormones.

A fantastic discovery of Italian physician and cytologist, Camillo Golgi, enabled him to view neurons under a microscope. A single neuron is so thin that light does not interact enough to make it visible under a microscope. Golgi invented a method to stain nerve tissue with silver nitrate, making them visible under the microscope. This opened the way to decipher the microscopic structure of nerve cells. The intricate drawings and analysis of these neurons by the Spanish histologist, Santiago Ramón y Cajal and his neuron theory led to the foundation of neurology. Cajal used the Golgi stain and developed a gold stain which allowed him to study the nerve cells in great detail. Cajal's studies established that neurons are not continuous and have gaps between them. It is because of this, neurons communicate through chemical transport across their walls.

The American physician and pharmacologist Otto Loewi discovered that the transmission of impulses from one nerve cell to another and from nerve cells to the intended organ involved chemicals. He and his colleagues identified the neurotransmitter acetylcholine, the first such chemical to be identified. Eventually, other neurotransmitters, such as serotonin, dopamine, noradrenaline, epinephrine, gamma amino butyric acid, enkephalin, *etc.* were discovered, and their role in our life started to untangle. Now we know that two nerves communicate through chemical signalling. If we interfere with the flow of these neurotransmitters in a neural pathway, we can achieve lots of things, many desirable and some undesirable. For example, anaesthetics block the sensation of pain during surgery. Some people use drugs like cocaine to alter their mood; it can make them abnormally happy, a state called euphoria. Consumption of hallucinogens like lysergic acid diethylamide (LSD), dimethyltryptamine (DMT) can result in our sensing something, which the rest of the crowd around us cannot see. When we drink whisky or wine, the ethyl alcohol in them interferes with the working of the brain to give a pleasurable sensation. Whatever we feel, hear, see around us is ultimately in the chemical coding in our nervous system. There are drugs like antidepressants, anti-anxiolytic, anti-psychotics that can help either to cure or to cope with mental diseases by modifying the working of the neurotransmitters.

Apart from neurological problems, which originate broadly in the malfunctioning of the nervous system, the brain suffers from chemical imbalances causing psychological errors. A major part of this psychology remains hidden in the unconscious. The great Austrian neurologist, Sigmund

Freud was suffering from mouth cancer, believed to have been caused by heavy smoking. Freud's death was the cause of an overdose of morphine administered by a doctor with the knowledge of Anna Freud, his daughter.

Freud evolved an entirely new understanding of the human mind. He created psychoanalysis, a clinical method to treat psychopathology. Freud postulated that the unconscious is central to the human mind; if the conscious mind is the tip of the iceberg, the portion below the surface is the unconscious. Freud delved into the innermost recesses of the human mind and proposed that the human psyche consists of the Id, the ego and the super-ego, an alternative to his previously proposed constituents of the conscious, preconscious and the subconscious.

We are yet to fully understand the working and mysteries of the brain; its malfunction can cause severe depression, paranoia, and anxiety, as in the case of Kurt Friedrich Gödel. The brain circuitry is also the reason for a person to be an enigmatic genius and of a completely different logical perspective, as was Kurt Friedrich Gödel.

LOGIC KILLS LOGIC

A mathematician, logician and philosopher, Gödel's brilliant insight into mathematical philosophy and logic had put "truth" to an existential crisis. Unfortunately, we cannot escape this truth; it is a difficult theorem, Gödel's incompleteness theorem.

The German mathematician, Georg Ferdinand Ludwig Philipp Cantor, proposed the concept of set and created the set theory that has become a fundamental theory in mathematics. He proposed that a set is a collection of distinct elements, where the elements can be anything as well as other sets. Philosophical insights from the set theory were so vast that mathematicians

> *Georg Cantor suffered from depression which worsened due to criticism of his works. He was sent to sanatoria on several occasions, finally dying of a heart attack in a sanatorium.*
>
> *Kurt Gödel suffered bouts of mental instability and had an obsessive fear of being poisoned. He would only consume food prepared by his wife. When she was in the hospital for six months, Gödel refused food and ultimately died of malnutrition! At the time of his death, Gödel weighed only 29 kg!*

could conceptualize even the abstract things of nature using the theory. When people started defining everything on the basis of set theory, the British polymath Bertrand Arthur William Russell and independently the German mathematician and logician Ernst Friedrich Ferdinand Zermelo discovered a flaw in the set theory. What happens, for example when we define a set of all the sets that do not contain themselves? Such a set does not contain itself and thus becomes eligible to be an element of the defined set. But if it becomes an element of the defined set, it contains itself! A paradox; if the set does not contain itself, it has to contain itself according to the definition! This is the Russel's paradox.

The contradiction was resolved by Zermelo and the Israeli mathematician Abraham Fraenkel when they developed a new set theory, the Zermelo-Fraenkel set theory, based on axioms which are fundamental mathematical rules. The American polymath John Von Neumann further extended the set theory, which avoided the contradictions of the previous systems. The Swiss mathematician Paul Isaac Bernays and Kurt Gödel made further fundamental additions and established the new set theory.

These sets can define everything; numbers, geometry, arithmetic operations, algebra in their own axiomatic systems. For example, the Italian mathematician

Giuseppe Peano's axioms of arithmetic. These five rules are what governs any arithmetic system with natural numbers.

We can say the value of the number of leaves in one tree is equal to the value of another tree after counting. This assessment of equality is the truth. We conclude it is the truth because we assume that the arithmetic rules of Peano's axioms are free of contradictions. However, Gödel constructed a theory that challenged our understanding of the world. Kurt Gödel proved that axiomatic systems, such as Peano's axioms, can prove something to be true. But for the axioms to be proven true, a set of new axioms are needed, and hence the axioms cannot prove their own consistency!

> *A Gödel machine is a hypothetical self-improving computer program that solves problems in an optimal way and can rewrite its own code. The machine was invented by Jürgen Schmidhuber, a computer scientist working in the area of Artificial Intelligence. The machine is named after Kurt Gödel, whose mathematical theories inspired it.*

We have had a glimpse of the working of our mind; it works on the biological principles of our brain. Biology works on the chemical bonds and reactions of molecules like neurotransmitters and proteins. The speed of those reactions and the exchange of energy is determined by fundamental laws of physics, like thermodynamics and electrodynamics. These theories are established with building blocks of mathematics, differential equations and algebra. These mathematical operations, when broken down, give us fundamental logical systems. Everything we have deliberated on so far, everything we think logical, everything we believe can be proven, is not true, and neither are they false. It is the limitation of science and philosophy and our mind; And From Here On, we hand over to God ...

www.ingramcontent.com/pod-product-compliance
Lightning Source LLC
LaVergne TN
LVHW050405160726
843469LV00041B/951